Oilwell Fishing Operations:
Tools and Techniques

Second Edition

Oilwell Fishing Operations: Tools and Techniques

Second Edition

Gore Kemp

Gulf Publishing Company
Houston, London, Paris, Zurich, Tokyo

Oilwell Fishing Operations:
Tools and Techniques

Second Edition

Copyright © 1986, 1990 by Gulf Publishing Company, Houston, Texas. All rights reserved. Printed in the United States of America. This book, or parts thereof, may not be reproduced in any form without permission of the publisher.

10 9 8 7 6 5 4 3 2

Library of Congress Cataloging-in-Publication Data

Kemp, Gore.
    Oilwell fishing operations: tools and techniques/Gore Kemp.—2nd ed.
      p. cm.
    Includes bibliographical references (p.).
    ISBN 0-87201-627-7 (alk. paper)
    1. Oil wells—Equipment and supplies—Fishing. I. Title.
TN871.K397 1990
622'.3382—dc20   90-33725
                                                                                CIP

Gulf Publishing Company
Book Division
P.O. Box 2608 ◻ Houston, Texas 77252-2608

Printed on Acid Free Paper (∞)

Contents

Foreword vii
Preface viii
Dedication ix
Chapter 1 An Overview 1
    Economics of Fishing.
Chapter 2 Avoiding Hazards 3
Chapter 3 Cardinal Rules in Fishing 5
    Dimensions. Leave Free Pipe Above Fish. Don't Rotate Fishing String. Look at Bottom of Pipe. Don't Pull Out of the Rope Socket.
Chapter 4 Pipe Sticking 7
    Common Causes. Example 1. Example 2. Solutions to Pipe-Sticking Problems.
Chapter 5 Determining Stuck Point 14
    Measuring Stretch. Buoyancy. Free-Point Instrument. Stuck-Pipe Logs.
Chapter 6 Parting the Pipe String 21
    Back-Off. Outside Back-Off. Chemical Cut. Jet Cut. Mechanical Cut.
Chapter 7 Catching Tools 29
    Overshots. Spears.
Chapter 8 Jarring Stuck Pipe or Tools 35
    Bumper Jar. Oil Jar. Jar Intensifier or Accelerator. Jarring Strings. Surface Jar. Drilling Jar.

Chapter 9 Washover Operations _____ 45
Washover Pipe. Rotary Shoes. External Cutters. Washpipe Spears. Unlatching Joint. Back-Off Connector. Hydraulic Clean-Out Tools.

Chapter 10 Loose Junk Fishing _____ 57
Magnets. Junk Baskets. Hydrostatic Bailer. Rotating Bailers. Junk Shots.

Chapter 11 Tungsten Carbide Mills and Rotary Shoes _____ 68
Material. Manufacture or "Dressing." Design. Running Carbide Tools.

Chapter 12 Wireline Fishing _____ 77
Cable-Guide Method. Side-Door Overshot. Radioactive Sources. Box Taps. Cutting the Line. Electric Submergible Pumps.

Chapter 13 Retrieving Stuck Packers _____ 88
Retrievable Packers. Permanent Packers.

Chapter 14 Fishing Coiled Tubing _____ 91

Chapter 15 Fishing in Cavities _____ 92
Bent Joints. Knuckle Joints. Induction Logs.

Chapter 16 Sidetracking Junk _____ 96

Chapter 17 Section Mills _____ 98

Chapter 18 Repair of Casing Failures _____ 100
Casing Leaks. Casing Back-Off. Stressed Steel Liner Casing Patch.

Chapter 19 Collapsed Casing _____ 105

Chapter 20 Fishing in High Angle Deviated and Horizontal Wells _____ 108

Chapter 21 Miscellaneous Tools _____ 109
Mouse Traps. Reversing Tools. Ditch Magnets. Mud Motors. Impression Blocks. Hydraulic Pull Tools. Tapered and Box Taps. Marine Cutting Tools.

Glossary _____ 117
Bibliography _____ 121
Index _____ 122

Foreword

Mr. Gore Kemp is one of the world's foremost experts on downhole problems in drilling or workover operations. His experience spans more than thirty years in all phases of operations dedicated to fishing tools and practices. He started Davis-Kemp, one of the first companies to specialize in fishing operations, and has passed along his vast storehouse of knowledge to others through seminars and personal contacts worldwide.

This book is an additional step by Mr. Kemp to share his knowledge with others. Many of the industry's experts fail to recognize their obligations to give back to the industry a small part of all the things the industry gave to them. We have all gained by Mr. Kemp's decision to write this book. His years of experience and expertise can now be utilized by others. For this we all owe Mr. Kemp a vote of thanks.

Preston L. Moore, Ph.D.
Norman, Oklahoma

Preface

A fishing job is an unwelcome but often necessary procedure in both drilling and workover operations. It is expensive. It usually is not in the budget, and the operator must see that it is performed in the most expeditious manner.

I have not found any reference that combines descriptions of the available tools and operating procedures as well as cautions and tips. Persons who are directly responsible for the operation need unbiased information on which to base their decisions. This book can serve as a text and reference for foremen, engineers, and superintendents who are writing procedures, making decisions, and supervising the operations.

The book gives descriptions and applications of fishing tools available in the industry along with do's and don't's based on experience.

In this updated second edition I have included information on fishing in high angle deviated and horizontal wells. Also, a new family of tools, the rotating bailers, reflects new concepts in downhole remedial and completion tools. The rotating bailers help prevent formation damage and fluid loss, and their use can lead to savings in cost and time. Since the publication of the first edition it has become possible to analyze fishing jobs by computer, and there are programs available to aid in the design of jarring strings.

In more than thirty years in the fishing tool business, I have observed thousands of jobs using various tools and procedures, and I have been exposed to the opinions and experience of several thousand drilling and production people through schools and seminars. This book represents the work of hundreds of operating people and I am indebted to them all.

Gore Kemp
Kilgore, Texas

Dedicated
to
The Glory of God

Chapter 1
An Overview

"Fishing" is the term used for procedures to correct downhole problems in an oil or gas well such as stuck pipe or drill collars, recovery of pipe twisted off or otherwise lost downhole, removal of loose junk, and the recovery or removal of wireline that has parted or become stuck. When any of these conditions develops, all progress in the drilling, workover, or completion ceases and fishing operations must be successfully completed before normal operations can resume.

Fishing is not considered to be a usual or common practice, but it is probably required to some degree in about one of every five wells drilled and up to four out of five wells that are worked over. Since the cost of fishing, including the rig time used, can be considerable, care and judgment must be exercised. Fishing tools and practices have been developed over the years making possible the correction of almost any downhole problem. However, the cost may be prohibitive, and in some cases, even initial fishing operations should not be conducted. In view of the high cost of rig operation plus the cost of the special services involved in fishing, proper judgment must be exercised and decisions must be made based on all the information available.

Fishing is not an exact science, and many times there is more than one way to approach the problem. However, there is probably a best way if all factors are considered. Personnel of fishing tool companies have valuable experience gained by performing this work constantly, where operating personnel are only exposed to these problems occasionally. Planning a fishing job is one of the most important phases, and costs can be reduced by adequate planning. Discussions should be held with all personnel involved, such as fishing tool operators or supervisors, mud company personnel, rig personnel, electric wireline company representatives (where applicable), and any others who might become involved. It is much cheaper to discover that a certain procedure will not work before doing it than after a misrun with the subsequent expense.

Economics of Fishing

Fishing should be an economical solution to the problem in the well. Obviously, a shallow hole with little rig time and equipment invested can justify only the cheapest fishing. When there is a large investment in the hole and substantial capital equipment to be recovered, more time and expense can be feasibly committed. There are studies, papers, formulas, and models that help in the economic decision of "to fish or not to fish, and if so, for how long?" All have merit, but so many factors affect the decision that converting them into a standard formula or pattern is almost impossible.

Probability factors are useful in determining the time to be spent on a fishing job. These percentages must be derived from similar situations, however, as there are no two fishing jobs exactly alike. Decision trees with the associated costs should be established for drilling and workover programs where there are multiple wells and similar situations.

Good judgment, a careful analysis of the problem, and then the skilled application of the decision insofar as the rig and tools are concerned is the best solution.

Chapter 2
Avoiding Hazards

There are many causes that contribute to a fishing job on both drilling and workover jobs but the predominant one is "human error." Many people in the industry feel that the majority of fishing jobs are man made. Certainly human error causes many fishing jobs to be done, but it should not be allowed to increase the time or expense of the fishing job itself.

There are some basic rules which should be followed during all drilling and workover operations that become even more important when fishing. Every effort should be made to recover something or to improve the situation on each trip in the hole with the tools. Misruns waste money and there is the possibility of additional mishaps on every trip in the hole. Probability indicates that a problem will develop during a given number of trips with the pipe.

Drawings noting dimensions should always be made of everything that is run in the wellbore. This responsibility should not be left to the service company personnel alone, but operating company personnel should also make independent measurements and sketches. If there is a large or unusual tool or downhole assembly being run, then a plan should be formulated as to how it would be fished if it should become stuck or broken. Ask "Can this tool be fished? Can it be washed over? And if so, what size washpipe can be run?" Keeping track of accurate dimensions of all equipment is a necessity if economical fishing is to be done.

Jars are frequently run as insurance against sticking. If there is a reasonable chance that the tool or assembly may get stuck, then jars run in the string are appropriate and the costs are probably justified.

Mud and other well fluids should be conditioned and have the desired properties prior to trips in the hole with fishing tools. It may be necessary to make a trip with a bit to condition the hole and circulate out fill that has covered up the fish.

When fishing, consideration should be given to releasing or recovering the fishing tools themselves should they become stuck or the fish cannot be pulled and the tool cannot be released. Ensure that the fishing tool works properly with the fish in question on the surface before running the tool downhole. If it does not perform properly on the surface, it is doubtful that it will be successful if run downhole.

Oil and gas wells represent tremendous investments. These can be lost quickly by carelessness or neglecting hazards that are always present.

Consideration should be given to the manner and speed with which the work string is run on fishing jobs. Care is taken on drilling jobs to not fracture the formation by lowering the drill string too fast and creating abnormal forces downhole. Care is also given to the pulling of the drill pipe so that the well is not swabbed and a blowout created. This same consideration should be given in both cased and open holes when fishing, because the fast pulling of the work string causes reciprocation at the bottom of the string and can cause the fish to be lost. This reciprocation or "yo-yoing" with a slight turning of the pipe can cause the jay in a packer retriever to release.

When fishing retrievable packers, it should be noted that the sealing element does not return to normal size for several hours. This close tolerance can cause problems in swabbing the wellbore, as well as hanging up in casing couplings if there is junk on top of the packer.

Oil and gas wells represent tremendous investments. These can be lost quickly by carelessness or neglecting hazards that are always present.

Chapter 3
Cardinal Rules in Fishing

Dimensions

Everything run in the well should always be measured and a dimensional sketch made for the record. Dimensions of internal bores should be noted, as some equipment and tools such as jars have restricted bores. Wireline tools should always be measured as they are not necessarily "standard."

Correct dimensions are most important for successful fishing. Overshots, for example, in sizes approximately 7-in. diameter and larger will usually catch a range of 3/16 inch. Smaller sizes have less range.

Many tools and instruments are run through the drill pipe, tubing, or fishing string. Dimensions of internal upsets or other restrictions are very important because tools, balls, or instruments can become lodged, plug the string, and possibly cause the entire operation to fail.

Equipment that is too large to fish in wellbores should be run only after careful consideration. It is not wise to say that this equipment should *never* be run, but the risk and the economics of the situation should be weighed properly.

Leave Free Pipe Above Fish

When parting a stuck string of drill pipe or tubing by the back-off method or by cutting, always leave some free pipe above the stuck point as a guide.

In circumstances where there is no fill to accumulate, one half joint may be adequate. When solids are falling out of the well fluid, it may be advisable to leave up to two joints above the then apparent stuck point. A second joint also provides extra insurance for catching or screwing into the fish if the top connection is damaged by the back-off shot or cutter.

In washover operations, free pipe is also left when backing off or cutting the fish. This provides a guide for the washpipe during subsequent washovers.

Leaving excessive pipe above the stuck point is not economical or good practice as the free fish is elastic in a jarring situation and requires additional pipe to cover in a washover job.

Don't Rotate Fishing String

To speed up a trip with a drilling or work string, the pipe in the hole is frequently rotated to unscrew the connection. During fishing operations, this is not an acceptable practice as the fish may be lost. The spinning of fishing tools such as an overshot, spear, magnet, junk basket, or washover pipe frequently causes the fish to be lost.

Look at Bottom of Pipe

Always inspect the bottom of the pipe that is removed from the wellbore. This practice should include not only all trips when fishing, but also those times when the pipe is parted for other reasons. In cases of twist-off, or failure of the pipe for any reason, the dimensions and configuration of the bottom of the parted joint will give excellent information for fishing. A good example of this practice is the inspection of the parted joint when pipe is jet cut. There is a flare that can be measured on the recovered piece, giving a dimension for the decision to mill with a milling tool or the overshot control.

On some occasions, the coupling on backed-off tubing will come out of the well on the last joint. Ordinarily it remains in the well because, having been made up at the mill, it is usually tighter. However, there are exceptions.

Don't Pull Out of the Rope Socket

Most conductor lines are connected to the tools or instruments with a rope socket shear device for a given pull. For some tools this means of retrieving can be acceptable, but it is a dangerous practice, particularly in open holes when running tools with radioactive sources. The most acceptable method of retrieving these tools is by cutting the line at the surface and stripping over (see page 77). If the line is pulled out at the rope socket, the instrument must be retrieved with a catching tool that may rupture the canister, allowing the radioactive material to pollute the well fluid. This can be an expensive and dangerous situation.

The fishing of conductor line or swab line with another line is not an acceptable practice. Fishing of these lines should be done with a pipe work string. Wireline can ball up and require considerable pull to retrieve it. Note that this comment does not include solid or "slick" line.

Chapter 4
Pipe Sticking

Common Causes

There are many causes of pipe sticking downhole and it is frequently desirable to identify the type of sticking so that the most effective method of recovery may be used. Some common types of pipe sticking follow.

Mechanical Sticking

Pipe may be mechanically stuck by packers, anchor-catchers, junk lost in the hole, multiple strings which have wrapped around each other, and crooked pipe that has been dropped or corkscrewed. Frequently when casing collapses, the tubing is stuck in the collapsed section. Mechanical sticking is more prevalent in cased holes than in open holes.

Mud Sticking

This can occur in both cased and open holes. It is usually caused by the settling out of solids in the mud, which is sometimes caused by high temperature setting up the mud. Casing leaks can allow shale and mud to enter the casing and stick the tubing or other equipment. Cuttings produced when drilling a well must be circulated out sufficiently to keep the hole clean; otherwise they will accumulate and cause sticking. Insufficient mud systems are frequently the cause of sticking in drilling wells. In some cases, wells have been drilled with clear water, and any mud used is that which is produced by the cuttings. This "native mud" can cause sudden sticking over a long interval and create a disastrous situation.

Key Seat Sticking

When a well deviates from the vertical, the subsequent rotation of the pipe and particularly the hard banded tool joints in the area of the "dog leg" wear a slot in the well bore that is smaller than the gauge hole (Figure 4-1). This undersize slot creates a hazard in "tripping" the pipe in and out of the hole. Frequently when pulling the pipe out of the hole, the larger drill collars are pulled up into this key seat and stuck. There is a natural tendency on the part of a driller to pull harder as he observes the pipe tending to stick. This, of course, merely makes the situation worse.

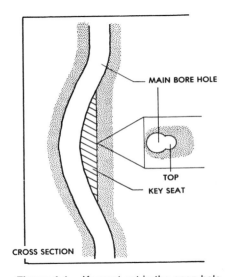

Figure 4-1. Keyseat cut in the open hole.

Cement Sticking

Cement sticking can occur due to a mechanical failure in equipment, a leak, human error, or intentional cementing in an attempt to contain a blowout or correct lost circulation. Many times when cement sticking occurs, premature or flash setting is blamed. The cuttings produced in drilling cement will readily stick the pipe if they are allowed to settle out of the fluid.

Blowout Sticking

When formation pressure exceeds the hydrostatic pressure of the mud or other well fluid, it causes shale, sand, mud, or other formation materials, and in some cases, even drill pipe protector rubbers to be blown up the hole, which sometimes bridges over and sticks the pipe.

Sloughing Hole Sticking

There is a tendency for shale sections to absorb water from the mud. These sections in turn swell and break off into the hole, lodging around the tool joints, drill collars, or the bit, causing the drill string to become stuck.

Undergauge Hole Sticking

A bit that has become worn under size by an abrasive formation may create this problem. It may be caused, however, by the formation expanding because of such things as salt flow, shale deforming, or the swelling of clay.

Lost Circulation Sticking

This very common problem occurs in formations ranging from shallow unconsolidated sands to formations that may be fractured by the excessive mud weights used. Lost circulation must be controlled by the use of the proper drilling fluid even after the drill string has become stuck and is being washed over.

Differential Pipe Sticking

This is one of the least understood causes of pipe sticking. It is caused by a high hydrostatic pressure creating a differential force that holds the pipe in a thick filter cake across a permeable zone (Figure 4-2). This situation may also become a very expensive and time-consuming problem.

Since differential sticking problems are usually solved by a variety of methods that are not applicable in any other type of pipe sticking, these methods will be discussed here prior to the usual jarring or washing over as discussed in Chapters 8 and 9.

Differential sticking occurs only across a permeable zone, such as sand, and the friction resistance may be a function of the filter cake thickness. The extra force necessary to pull the pipe loose from the wall may be calculated by the following formula:

$F = DP \times A_c \times C_f$

where F = force in pounds
DP = differential pressure in psi
A_c = area in contact in sq in.
C_f = coefficient of friction

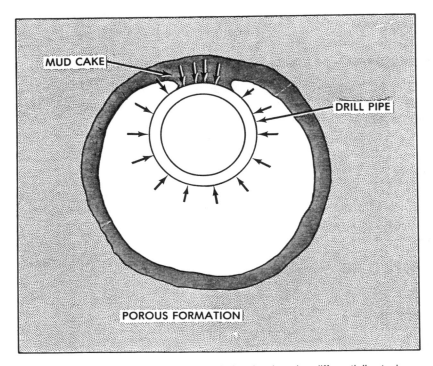

Figure 4-2. Cross section of an open hole, showing pipe differentially stuck.

It can be readily seen by calculating the forces in two hypothetical situations that the pull necessary to free the pipe frequently far exceeds the tensile strength of any pipe available.

Example 1

Assume that drill pipe contacts the filter cake in a width of 3 in. along a 25 ft sand zone with a pressure differential of 1,600 lb/in.^{2} and a friction coefficient of 0.2.

F = 1,600 psi × (25 ft × 12 in./ft × 3 ft) × .2

F = 28,000 lb

This force must be added to the normal hook load to pull this pipe free. In many cases the total load would exceed the safe pull on the pipe.

Example 2

Assume that 7 in. drill collars are stuck in a 9⅞ in. hole with a high water-loss mud. It is quite possible that shortly after the drill collars become stuck that one-third of the circumference of the drill collars is imbedded in the thick filter cake leaving two-thirds of the drill collars exposed to the hydrostatic pressure of the mud column. This could be calculated in an 11,000-ft hole as follows:

$$F = HP - PP \times 1 \text{ ft} \times 12 \text{ in./ft} \times \tfrac{1}{3}C \times C_f$$

where HP = hydrostatic pressure
  PP = pore pressure in lb/in.^{2}
  C = circumference of 7-in. O.D. drill collars
  C_f = arbitrary coefficient of friction = .2

Thus

$$F = (8,294 - 8,120) \times 1 \text{ ft} \times 12 \text{ in./ft} \times 7.3304 \times .2$$

$$F = 3,061 \text{ lb/ft of stuck drill collars}$$

If one 30 ft drill collar is stuck, then an additional force of 91,830 lb is necessary to pull free.

The causes of differential sticking may be listed as follows:

1. High ratio of wellbore pressure to formation pressure.
2. Large drill collars in relation to the hole size.
3. High filtration rate.
4. High mud-solids content.
5. Excessive shutdown time opposite a permeable formation.

Normally the sticking occurs when the drill pipe is not in motion, and usually full or partial circulation can be accomplished. The immediate step to be taken is to shut down the pumps. Pump pressure during circulation increases the wellbore pressure slightly. Stopping this additional pressure may be enough to reduce the force sufficiently that the pipe may be worked free.

Solutions to Pipe-Sticking Problems

Surge Method

The surge or U-tube method of freeing the stuck pipe involves displacing a portion of the mud system in the hole with a lighter weight fluid and allowing the system to flow back to a balanced position. This lighter fluid may be diesel oil, crude oil, water, nitrogen, gas, or any fluid that is available with an appropriate weight. The quicker this can be accomplished, the more effective it will be. When the fluid is flowed back, the fluid level in the annulus is lower, therefore the hydrostatic pressure on the formation is reduced. If this is sufficient to at least equal the formation pressure, the string will come free. This method of freeing the pipe is safe since the pressure can be reduced in several steps so that a dangerous situation can be avoided. The mud weight itself is not reduced, and if a kick occurs it can be controlled by the fluid which was flowed out of the annulus. Since the displacing fluid is all contained in the drill pipe, there is a minimum of dilution and the filter cake in the hole is not affected.

In some cases, particularly where gas is used, displacement is done down the annulus and the heavier mud flowed out the drill pipe. This must be done very carefully and slowly since the pressure necessary to flow the mud back may break down the formation.

In each case, calculations should be made for displacements and only those amounts necessary to free the pipe used. This is, however, a very effective method used in some areas with a high rate of success.

Spotting Fluid

If there is not sufficient reduction in pressure to free the pipe, then usually it is advisable to spot a fluid across the stuck zone which will penetrate the filter cake and remove it. The fluid used depends on the formation and the composition of the mud cake. Surfactants are most useful in these spotting fluids, as they reduce the interfacial tension between the contacting surfaces. A great deal of work and research has gone into the most appropriate material for the dissolving of the filter cake. Chemicals that penetrate and crack the cake have been very successful. There have been several patents issued on combinations of materials to be used for this purpose.

Diesel and crude oils are used most commonly with the proper surfactant in the mixture. The most usual problem with this method of freeing the pipe is that the operator will not spend enough time to allow the filter cake to be removed. The freeing fluid is invariably lighter than the mud in the hole so there is going to be considerable migration up the hole after it is spotted. It is necessary that a new slug be spotted about every thirty minutes. All of this is done after the displacement of the drill string and the hole up to the stuck area is calculated. Most studies of freeing pipe by this method indicate that at least eight hours should be allowed for the procedure to take effect. It is not advisable to pull on the pipe during this time, as it will merely pull down into the wellbore even more. A small weight should be left resting on the stuck portion so that it is known when the pipe becomes free. This statement is controversial, and some believe that the pipe should be worked constantly. This is allowable if very small increments of total pipe weight are used. Torquing the pipe during this time is advisable, however; and small amounts of weight can be left on the stuck pipe if it is off bottom.

Drill Stem Test Tool

This is one method of freeing differentially stuck pipe used most effectively but which has not been universally accepted because of other inherent hazards of the operation. Open-hole packers or test tools may be used to remove the hydrostatic force from the stuck pipe and to free it the instant the tool is set.

The purpose of the DST tool is to lower the hydrostatic pressure around the fish enough to allow the formation pressure to push the fish away from the wall. The fishing string consists of a catching tool or screw-in sub on bottom, a perforated sub in case the fish is plugged, bumper jars, packer and optional safety joint, and jars above the test tool.

A packer seat that will support the tool and the weight of the mud column above the tool must be selected. By backing off the pipe string and spacing out the fishing string the test tool will be located in the appropriate zone.

To operate the tool, the string is run and the fish caught or screwed in. The weight of the string is set down on top of the fish which causes the packer to expand and seal off. This separates the mud column above the packer from the hole below, greatly reducing the hydrostatic head in the stuck section. As weight is applied to the string a bypass valve is closed and a valve opened so that the pressure trapped below the packer escapes into the drill string. The pressure in the formation immediately pushes the stuck pipe away from the wellbore. As the string is picked up, the packer unseats and contracts, the connecting valve closes and the bypass valve opens. The fish may then be pulled from the wellbore.

If none of the preceding methods is successful, it will be necessary to part the pipe and either jar on it or wash over. Ordinarily jars are used if the stuck interval is short. If there is a great deal of pipe to be freed, most operators will wash over. Each of these operations is discussed in subsequent chapters.

After the cause of the pipe sticking has been determined, plans must be made to free and recover the pipe. Some of the fishing procedures recommended for the particular problems follow.

Chapter 5
Determining Stuck Point

Measuring Stretch

When pipe becomes stuck for any of the reasons described, the first step is to determine at what depth the sticking has occurred.

Stretch in pipe can be measured and a calculation made to estimate the depth to the top of the stuck pipe. All pipe is elastic and all formulae and charts are based on the modulus of elasticity of steel, which is approximately 30,000,000 lb/sq in. If the length of stretch in the pipe with a given pull is measured, the amount of free pipe can be calculated or determined from a chart available in data books.

Since all wellbores are crooked to some extent, there is friction between the pipe and the wellbore. Steps should be taken to reduce this friction to a minimum. The pipe should be worked for a period of time by pulling approximately 10%–15% more than the weight of the string and then slacking off an equal amount.

There are certain techniques that reduce error in estimating stuck points from stretch data. It is also necessary to assume certain arbitrary conditions. Stretch charts and formulas do not take into consideration drill collars or heavy weight drill pipe.

First, pull tension on the pipe at least equal to the normal hook load (air weight) of the pipe prior to getting stuck. This should then be marked on the pipe as point "a." Next, pull additional tension which has been predetermined within the range of safe tensional limits on the pipe. Now slack off this weight back down to the hook load weight. Mark this point "b." It will be lower than point "a." This difference is accounted for by friction of the pipe in the wellbore.

Next pull additional tension on the pipe to a predetermined amount within the safe working limits of the string. Mark this point as "c." Pull additional tension on the pipe in the same amount used to determine points "a" and "b" and slack off to the tension used to locate point "c."

The mid-point between "a" and "b" and between "c" and "d" will be the marks used. Measure the distance between these average marks and use this number as the stretch in inches.

The amount of free pipe can be determined by using the following formula:

$$\text{length free pipe (ft)} = \frac{1{,}000{,}000 \text{ (stretch-in.)}}{K \text{ (pull over wt. string)}}$$

where K = constant

The constant in this formula can be determined by:

$\dfrac{1.5}{\text{wt lb/ft}}$ for drill pipe or $\dfrac{1.4}{\text{wt lb/ft}}$ for tubing and casing.

This method of estimating the stuck point of pipe is not completely reliable and accurate as there are too many variables caused by friction, doglegs, hole angle, and pipe wear. However, it frequently indicates the cause of sticking such as key seats or differential sticking in open holes and collapsed casing or casing leaks in producing wells.

Rather than calculating the stuck point, there are two types of stretch charts found in many data books from which the length of free pipe can be read directly.

The nomograph type charts (Figure 5-1) consist of three columns of numbers. The first two columns are the pull (lb) and the stretch of pipe (in.), both of which are known; the third column gives the free length of pipe (ft), which is the unknown. By laying a straight edge across the two known numbers, one can read directly the unknown depth at which the pipe is stuck.

The straight line curve chart (Figure 5-2) is a graph on which the stretch in inches is laid off on the horizontal axis and the unknown depth to the sticking point in feet is laid off on the vertical axis. The pull in pounds in excess of the weight of the pipe string is expressed as a straight line drawn at an angle between the two axes.

Either type of chart will give an approximation of the depth at which the pipe is stuck, and this information in many cases will indicate the reason for the sticking or at least rule out other causes.

Accuracy of the charts and the formula is approximately the same, as both are affected by the same problems of hole friction, loss of material in used pipe, and the accuracy of weight indicators. Note, however, that the modulus of elasticity of all grades of steel is the same. The grade of the pipe does not affect its stretch.

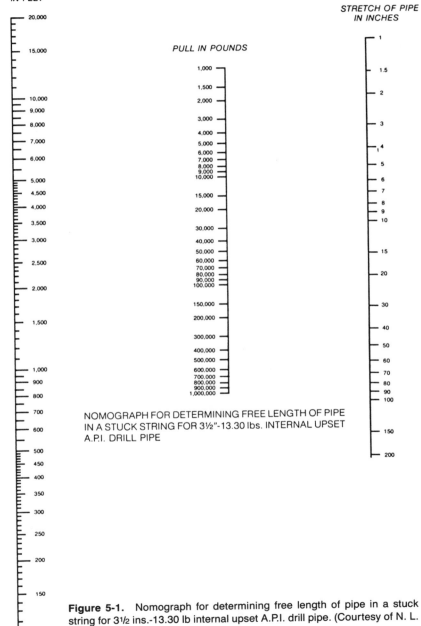

Figure 5-1. Nomograph for determining free length of pipe in a stuck string for 3½ ins.-13.30 lb internal upset A.P.I. drill pipe. (Courtesy of N. L. McCullough.)

Determining Stuck Point 17

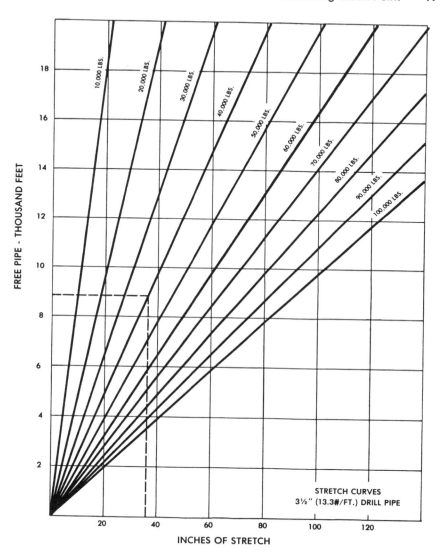

Figure 5-2. Straight-line stretch chart.

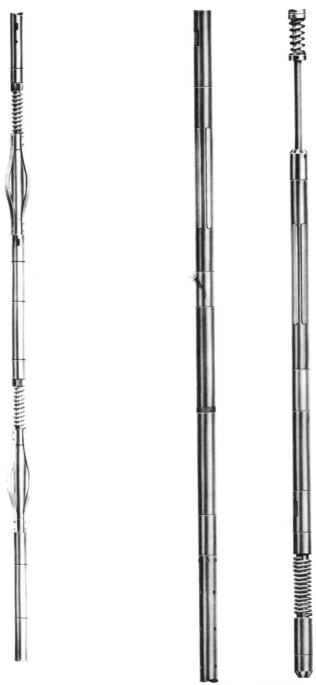

Figure 5-3. Free point indicator tool using springs for anchoring in pipe. (Courtesy of N. L. McCullough.)

Figure 5-4. Free point indicator tool using magnets for anchoring in pipe. (Courtesy of N. L. McCullough.)

Buoyancy

This force must be dealt with constantly in drilling wells and to a lesser degree in producing wells or cased holes. It may be a considerable factor in determining such variables as the number of drill collars to run. As an example, a drill collar has a buoyed weight of only approximately three-fourths of air weight in 16 ppg mud. However, when pipe is stuck, the buoyant forces are being exerted against the stuck section, and therefore there is no *effective* buoyant force at the surface. Immediately when the pipe is freed, the buoyant forces are again in effect and are to be reckoned with accordingly.

This statement is, of course, ignoring the cumulative length of the tool joints or couplings and the small hydrostatic forces tending to buoy them.

Free-Point Instrument

Electric wireline service companies run instruments on conductor lines and are able to accurately determine the stuck point of pipe. The instruments are highly sensitive electronic devices which measure both stretch and torque movement in a string of pipe. This information is transmitted through the electric conductor cable to a surface panel in the control unit where the operator interprets the data.

The basic free-point instrument (Figure 5-3) consists of a mandrel which encompasses a strain gauge or microcell. At the top and bottom of the instrument are friction springs, friction blocks, or magnets (Figure 5-4), which hold the tool rigidly in the pipe. When an upward pull or torque is applied at the surface, the pipe above the stuck point stretches or twists. The change in the current passing through the instrument is measured by the microcell and transmitted to the surface for interpretation. When the instrument is run in stuck pipe, there is no movement of the pipe, therefore there is no pull or torque transmitted to the instrument. In turn, the gauge at the surface shows no change in its reading.

Free-point indicators are frequently run with collar locators and in combination with string shots, chemical cutters, and jet cutters. This combination run saves expensive rig time, and it will also maintain a continuous sequence in measuring so that there is less chance of a misrun in cutting or backing-off.

Since fishing operations are usually begun as soon as the pipe is parted following the free-point determination, it is a good practice to have the fishing tool supervisor or operator on the location during the free-point and back-off or cutting operations. The fishing tool operator needs to observe the free-point and parting operations, as frequently there are suggestions that can be made to improve the fishing situation.

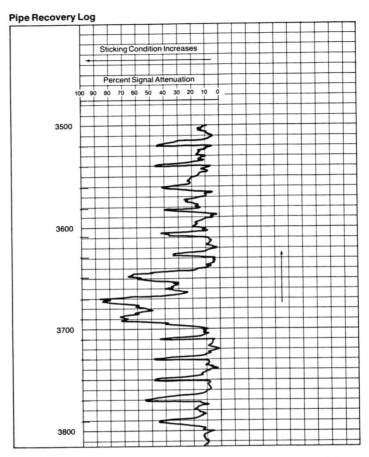

Figure 5-5. Pipe recovery log. (Courtesy of N. L. McCullough.)

Stuck-Pipe Logs

A log which measures the severity and the length of stuck pipe is very helpful in determining what method to use to free the pipe. Shown in Figure 5-5 is a pipe recovery log which expresses the sticking condition as a percentage. A vibration is used and measured by a receiver. At stuck intervals, the sonic vibrations decrease in proportion to the severity of the sticking. The downhole instrument is calibrated in known free pipe; normally near the bottom of the surface pipe. The pipe recovery log gives a complete record of all stuck intervals and possible trouble areas in a string of stuck pipe. This information is very helpful in evaluating conditions to determine whether to jar on the stuck section, to wash over the fish, or in some cases, to sidetrack. It may be used in drill pipe, tubing, casing, or washpipe.

Chapter 6
Parting the Pipe String

After determination of the stuck point in a pipe string, it is normal procedure to part the string so that fishing tools such as a jarring string or a washpipe string may be run.

There are four acceptable methods of parting the pipe string:

1. Back-off—Unscrewing the pipe at a selective threaded joint above the stuck point using a prima cord explosive run on an electric wireline.
2. Chemical cut—An electrical wireline tool and procedure that uses a propellant and a chemical, halogen fluoride, to burn a series of holes in the pipe thereby weakening it so that it easily pulls apart with a slight pull.
3. Jet-cut—A cut made by an explosive shaped with a concave face and formed in a circle. It is also run and fired on an electric line.
4. Mechanical cut—A cut made with a set of knives installed in a tool and run on a small-diameter work string. This is referred to as an internal or inside cut. The pipe may also be cut with an external cutter run in conjunction with washpipe. Short sections of pipe which have been washed over are cut by this outside method. The inside mechanical cut has been replaced to a large degree by the chemical and jet cutters because they are run on conductor lines, while the mechanical cutter must be run on small pipe or sucker rods. This requires picking up the pipe and making a trip with the tool. Rig time is the operator's most expensive cost, so in most cases tools run on electric line are more economical overall.

In the 1950s, pipe, both tubing and drill pipe, was usually parted by mechanical cutters, but with the advent of various methods of cutting pipe with tools run on electric wirelines, the speed with which the operation can be accomplished has been increased to the point where most operations are done on wirelines if possible. Research is continuing on various methods of cutting pipe and other material. The latest is a pyrotechnic cutting system which is nonexplosive, but acts as a cutting torch to instantaneously cut or perforate steel, including pipe. This method is in limited service at the present time, but it is expected to become more popular since it cuts with a pressurized flame jet leaving a smooth cut. The ignition of the cutting material is initiated by a high-voltage igniter, which makes it safe from outside sources of initiation thereby increasing the safety for personnel and equipment.

The cutting method for the particular job should be selected carefully. Only the back-off method leaves threads looking up, and therefore should be selected if a retrieving tool is to be screwed in the fish.

One rule that should always be followed when parting pipe is to leave free pipe above the stuck point as a guide and to ensure a catching surface long enough for good pulling strength. Sufficient length for these purposes is usually considered to be between one-half joint to two joints. Consider the operation to immediately follow when determining the amount of free pipe to leave. For example, if washing over inside casing and using a set-up where no threads are needed is to be done, cutting one-half joint above the stuck point may be the logical decision. Or if drill pipe is backed off so washing over (using a drill collar spear in the washpipe) can be commenced, then extra pipe is left in the hole if there is considerable settling out of solids. Some operators also like a "spare" tool joint in case the first is damaged by the back-off operation. Never leave more pipe above the stuck point than needed, however, since it contributes to more washing over if this method is used. If jarring is done extra pipe adds elastic pipe to the fish.

Back-Off

Back-off is the procedure of applying left-hand torque to a pipe string and firing a shot of prima cord explosive which produces a concussion to effectively partially unscrew the threads (Figure 6-1).

The back-off method of parting pipe is probably the most popular of all, particularly in drill pipe, since this is the only method which leaves a threaded connection up, making it possible to screw back into the fish with a work string including the fishing tools such as jars. This is important in the case of drill pipe since this method eliminates a "catching" tool such as an overshot. Frequently there is little clearance in the open hole and the elimination of large-diameter tools is desirable. Tool joints used on drill pipe, drill collars, and other drilling tools have coarse threads, large tapers and seal by the flat surfaces or faces. These characteristics make the back-off method attractive.

Tubing or other coupled pipe does not lend itself to back-off in the same way. Threads are usually fine, at least eight per inch; there is only a slight taper, such as ¾ in. per foot, and the threads are in tension with a high degree of thread interference. In spite of these differences, back-off is still a popular method of parting tubing. Usually after a back-off in a tubing string, an overshot is run as the chance of cross-threading the fine tubing threads is great.

To prevent accidental back-off in a loose connection up the hole, the pipe should first be tightened. This is accomplished by applying a specific number of rounds of right-hand torque and then reciprocating the pipe while holding the torque. By counting the rounds of torque make-up and then counting the rounds that "come back" when the tongs or rotary are released, rounds of make-up in the threads somewhere in the free pipe are indicated. Using API torque, unless there is some good reason to vary this amount, the procedure should be continued until there is no more make-up.

Once the pipe is made up, left-hand torque is introduced in the string. This torque must also be "worked down" by reciprocating the pipe as the torque is built up. This action distributes the torque throughout the string and assures that there is left-hand torque at the point of back-off. A well-accepted rule of thumb for the amount of torque applied to the pipe is one round per thousand feet for two-inch and two-and-one-half-inch tubing. Four-and-one-half-inch drill pipe would require only about half the rounds to build up sufficient torque.

Theoretically just prior to firing the string shot, the pipe at the back-off point should be in a neutral condition, with neither tension nor compression. Since this condition is very difficult to obtain, any choice should lead toward slight tension in the pipe. Since buoyancy is not effective in the stuck pipe, air weight is used in calculations. However, the moment the pipe starts to spin free, the flat horizontal face of the pipe is uncovered, allowing buoyancy to produce a lifting force against the string. This force is affected by the cross-sectional area of the tool joint face, the depth, and the mud weight. The left-hand torque is held, and the determined weight of the string is picked up when the string-shot is fired. The concussion at the joint momentarily loosens the threads and the pipe begins to unscrew. It usually must be manually unscrewed completely and then the freed pipe can be removed from the well.

When ordering a string-shot, the service company needs to know the size and weight of pipe to be backed off, the approximate depth of the stuck point, the weight of the mud or fluid in the hole, and the temperature of the well. This information will dictate the strength of the charge needed as well as the type of fuse.

String shots are also used for other purposes such as:

- Releasing stuck packers or fishing tools.
- Removing corrosion from pipe.

- Opening perforations.
- Jumping collars.
- Removing jet nozzles from drill bits to increase circulation.
- Knocking drill pipe out of key seats in hard formation.

Outside Back-Off

String shots can also be run in the annulus and pipe backed-off from the outside. When pipe is plugged and it is not possible or practical to clean it out so that a string shot can be run inside the pipe, it may be practical to run the string shot outside the pipe in the annulus. Usually a back-off is made internally as deep as is possible and the free string removed. A sub which has a side opening is made up on the bottom of the string to screw back in the fish (see Figure 6-2). When the free string is run and made up with the side-door sub on bottom and screwed into the fish, the conductor line and string shot are run inside the pipe down to the side-door sub where it is guided into the annulus and lowered deeper. To lower the electric line and the string shot in the annulus, the service company will rig up a very slim connection with a flexible flat weight that can be worked down through the small clearances. The back-off is accomplished in the same manner as the inside operation with left-hand torque and the pipe weight picked up. The side-door sub is also known as a "hillside sub" in some areas.

Chemical Cut

This method of cutting pipe is the most recent innovation. It was first used in the fifties. It was patented and for years was an exclusive process of one wireline company. Today it is available through most electric wireline service companies. All wireline cuts are economical because rig time is reduced to a minimum. The big advantage of the chemical cut (Figure 6-3) is that there is no flare, burr, or swelling of the pipe that is cut. No dressing of the cut is necessary in order to catch it on the outside with an overshot or on the inside with a spear.

The chemical cutting tool (Figure 6-4) consists of a body having a series of chemical flow jets spaced around the lower part of the tool. The tool contains a propellant which forces the chemical reactant through the jets under high pressure and at high temperature to react with the metal of the pipe. Electric current ignites the propellant which forces the chemical, halogen fluoride or bromine trifluoride, through the reaction section which heats the chemical and forces it out the jets. The tool also contains pressure-actuated slips to prevent a vertical movement of the tool up the hole, thereby fouling the electric line.

25

Figure 6-2. Side-door back-off sub.

Figure 6-1. String shot back-off tool. (Courtesy of N. L. McCullough.)

Figure 6-3. Pipe cut with a chemical cutter. (Courtesy of N. L. McCullough.)

Figure 6-4. Chemical cutter. (Courtesy of N. L. McCullough.)

The chemical cutting tool may also be explained as producing a series of perforations around the periphery of the pipe. The reaction of the chemical with the iron of the pipe produces harmless salts which do not damage adjacent casing. The products of the chemical reaction are harmless and are rapidly dissipated in the well fluid.

The chemical cutter will not operate successfully in dry pipe but requires at least one hundred feet of fluid above the tool when a cut is made. The fluid should be clean and contain no lost circulation material. The chemical cutter has operated successfully at a hydrostatic head pressure of 18,500 psi and 450°F. It is available for practically all sizes of tubing and drill pipe and most popular sizes of casing.

Since it is not necessary to torque up the pipe when chemically cutting as compared with string shot back-off, it is a safer process for personnel on the rig floor.

Jet Cut

The jet cutter (Figure 6-5) is a shaped charge of explosive which is run on an electric wireline. The modified parabola face of the plastic explosive is formed in a circular shape to conform to the shape and size of the pipe to be cut. When an explosive such as this is used to cut pipe, the end of the pipe is flared (Figure 6-6), and it is necessary to remove this flare if the pipe is to be fished with an overshot from the outside or from the inside with a spear. Usually this can be accomplished on the same trip with the retrieving tool.

A mill control or a mill container guide can be run with an overshot and the flare or burr removed by rotation so that the fishing tool can slip over the fish and catch it.

The jet cutter is often used when abandoning a well during salvage operations or when low fluid level, heavy mud, or cost would preclude the use of the chemical cutter.

There is a possibility of damage to an adjacent string or to casing if the pipe to be cut is touching at the point where the cut is made.

Jet cutters are available for practically all sizes of tubing, drill pipe, and casing. The same principle is used in special jet cutters for severing drill collars.

Mechanical Cut

The pipe string may also be parted by using a mechanical internal cutter. Ordinarily, to part the string in order to run fishing tools, the pipe is parted by wireline methods as rig time is held to a minimum. If, for some reason, wireline tools are not available or practical, the pipe may be parted by running an inside cutter on a string of small-diameter pipe or sucker rods. The time that is consumed in procuring the small string of pipe, picking it up, and running it usually makes this method a poor choice from an economic viewpoint.

Figure 6-5. Jet cutter. (Courtesy of N. L. McCullough.)

Figure 6-6. Pipe cut with jet cutter. (Courtesy of N. L. McCullough.)

Figure 6-7. Internal mechanical cutter. (Courtesy of Bowen Tools.)

The internal cutter (Figure 6-7) is made on a mandrel with a wickered sleeve or split nut fitted to threads on the mandrel and commonly called an "automatic bottom." This allows the slips to be released and the tool to set at any specific depth desired. Friction blocks or drag springs are fitted to the mandrel to furnish back-up for this release operation. As weight is applied to the set tool, knives are fed out on tapered blocks, and as the tool is rotated, they engage the pipe and cut it in two.

In practically all cutting tools, springs are provided in the feed mechanism to absorb any shock that may accidentally be applied to the work string causing the knives to gouge or to break.

Fishing tool operators will usually run a bumper sub above the cutter so that excessive weight is not exerted on the knives causing them to break or dig into the pipe. Sinker bars may also be used in the string to give the correct amount of weight while the bumper jar is run in the free-travel position.

Chapter 7
Catching Tools

Overshots

The overshot is the basic outside catch tool and is probably the most popular of all fishing tools. The style designed with the helical groove in the bowl and the grapple or slip made to fit this design is now almost universally used, and will therefore be the only one discussed here.

Most overshots consist of a bowl, top sub, guide and the grapple or slip, a control, packoff, stop, and perhaps some additional accessory. The overshot bowl is turned with a taper on a helical spiral internally and then the grapple, which is turned with an identical spiral and taper, is fitted to it. Each grapple is turned with a slip or wickered surface inside so that a firm catch is assured. Depending on the size of the catch for which it is designed, a grapple will be either the basket type (Figure 7-1) for relatively small catches, or the spiral type (Figure 7-2) for large catch fish in relation to the outside diameter of the bowl.

The type of grapple furnished should not concern the operator, as this is strictly a matter of size and manufacturing design. It is not possible to order either of the specific types if the relative size does not fall into that category. Since spiral grapples appear to be weak and even "flimsy" in some cases, many persons are concerned about their strength. In actual practice, the spiral grapple makes a stronger assembly because it is flexible and distributes the load throughout the bowl. Most overshots fail through overstressing, and it is then that the bowl splits or swells due to exceeding the design limitations.

An interesting comparison between the two designs of grapples is the capacity for a $7^5/_8$-in. full strength overshot fitted with a spiral grapple as compared to the capacity for the same overshot fitted with a basket grapple. The load capacity with the basket grapple is 479,044 lb, while it is 542,468 lb with a spiral grapple.

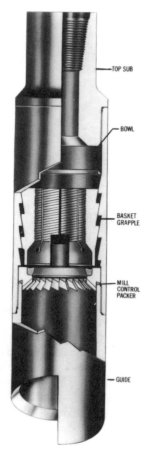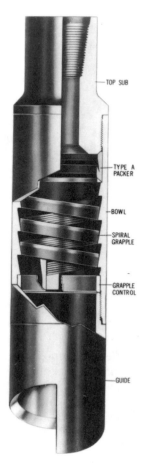

Figure 7-1. Overshot dressed with basket grapple. (Courtesy of Bowen Tools.)

Figure 7-2. Overshot dressed with spiral grapple. (Courtesy of Bowen Tools.)

Location of the grapple within the bowl is controlled by means of a cylindrical ring with a tang or key which fits into a slot and prevents the grapple from turning but allows it to move up and down on the tapered surface. Because of the design, the grapple contracts as it is pulled down on the tapered surface and grips the fish more firmly as the pull is increased.

Controls may also be designed with a pack-off, or packer, that seals off around the fish and allows the circulating fluid to be pumped through the fish. Ordinarily this helps to free the object if it is stuck.

Care must be exercised in fitting or "dressing" an overshot where a coupling or tool joint is to be caught. It is necessary that the enlarged section of the pipe to be caught (tool joint or coupling) be positioned in the grapple wickered area. If it moves up and above this section, the overshot may rotate freely, and it becomes impossible to release. Stops of various designs are used to stop the enlarged "catch" in the proper grapple area. Some of these are simple doughnut-shaped rings placed in the bowl above the grapple; others may incorporate a spring-loaded packer, or pack-off, to seal inside a tubing coupling while others consist of an internal shoulder at the top of the grapple itself.

Basket grapple mill control packers should always be run when fishing for drill pipe when the catch is small enough to accommodate a basket grapple. Frequently there are burrs, snags, and splinters on the pipe that is to be caught. The mill is sized so that it will trim these enlargements down to the proper size to be caught by the grapple. When the pipe has been "shot off" or parted in such a way to heavily damage it, it may be necessary to fit a mill extension, or mill guide, to the overshot bowl so that extensive milling can be accomplished for the catch to be made on the same trip in the hole. These extensions, or guides, are "dressed" inside with tungsten carbide and can mill off a substantial amount of material so that the "fish" is trimmed down to the grapple size.

Overshots are very versatile and may be fitted for almost any problem. Extensions such as washover pipe may be run above so that pipe can be swallowed and the overshot fitted to catch the coupling or tool joint below. This is recommended frequently when the top joint of the fish is in such bad condition that it is not practical to pull on it.

Short catch overshots (Figure 7-3) are available in limited sizes to use where the exposed portion of the fish is too short to be caught with a conventional catch overshot. The wickered or catching portion of the grapples in short catch overshots usually begins within one inch of the bottom of the bowl.

To release overshots designed in the foregoing manner, it is first necessary to free the two tapered surfaces, bowl, and grapple, from each other. In pulling on a fish, these two surfaces have been engaged and the friction would prevent release. This freeing of the grapple or "shucking" can be accomplished by jarring down with the fishing string. Usually a bumper jar or sub is run just above the overshot and is used for this purpose. Before jarring down, one should be sure that the oil jar is closed to prevent damage to the seals. After bumping down on the overshot the grapple is usually free and the overshot can be rotated to the right and released from the fish. Slight pull upward should be exerted on the overshot, as this pull plus the lead turned in the slip surface of the grapple will screw it off the fish. If a large amount of the fish has been swallowed, it may be necessary to free or "shuck" the grapple more than once.

To properly engage an overshot on a fish, slowly rotate the overshot as it is lowered onto the fish. The pump may be engaged to help clean the fish and also to indicate when the overshot goes over the object being caught. Once this has been indicated by an increase in pump pressure,

32 Oilwell Fishing Operations

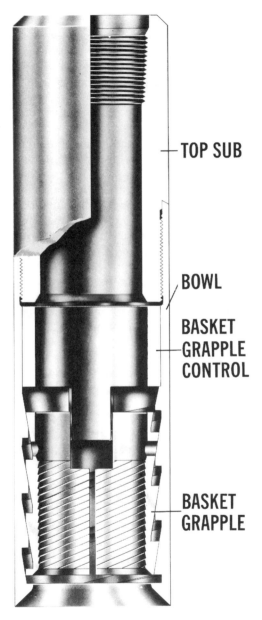

Figure 7-3. Short catch overshot. (Courtesy of Bowen Tools.)

Figure 7-4. Releasing spear. (Courtesy of Bowen Tools.)

stop the pump, as there may be a tendency to kick the overshot off the fish. An overshot should not be dropped over the fish. If jarring is to be done, it should be started with a light blow and gradually increased as this tends to "set" the grapple on the fish. A hard impact up may strip the grapple off the fish and cause the wickers to be dulled. This procedure can cause a misrun and a trip to replace the grapple.

Spears

Spears (Figure 7-4) are used to catch the inside of pipe or other tubular fish as opposed to overshots which catch on the outside. Usually a spear is not the first choice if there can be a choice between the two, as the spear has a small internal bore which limits the running of some tools and instruments through it for cutting, free-pointing, and in some cases, backing-off. The spear is also more difficult to pack off, or seal, between the fish and the work string than is an overshot.

However, spears are popular for use in pulling liners, picking up parted or stuck casing, or fishing any pipe that has become enlarged when parted due to explosive shots, fatigue, or splintering. Due to the design with the small bore in the mandrel, spears are usually very strong. For a comparison, one manufacturer produces a spear to pick up 5 1/2-in. casing with 4 1/2-in. drill pipe that has a strength of 628,000 lb. An overshot made for the same catch would have a strength of 580,000 lb. Obviously, either of these tools in this size is adequately strong since 4 1/2-in. 16.60 lb Grade S drill pipe has a yield strength of 595,000 lb and Grade E 330,000 lb.

The most popular spears in use today are built on the same principles as the overshots described earlier. They are designed with a tapered helix on the mandrel (as the tapered helix turned inside the bowl of the overshot) and a matching surface on the inside of the grapple. The slip, or gripping surface, of the grapple is on the outside surface of the spear so that it will catch and grip the inside of the pipe that is being fished.

In order to release a spear, it is rotated to the right (Figure 7-5). If the grapple is frozen to the mandrel, it may be necessary to bump down to free or "shuck" the grapple. Usually a bumper jar or sub is run just above the spear and this can be used to effectively jar down and free the grapple. To prevent damage to the seals in any oil jar that is run in the string, the oil jar should be closed before jarring down.

The spear is a very versatile tool, in that it can be run in the string above an internal cutting tool or in combination with other tools, thereby saving a trip in the hole with the work string. Milling tools may be run below the spear to open up the pipe so that the spear can enter and catch.

34   Oilwell Fishing Operations

Figure 7-6. Spear pack-off. (Courtesy of Bowen Tools.)

Figure 7-5. Releasing spear: (A) Catching position; (B) released position. (Courtesy of Bowen Tools.)

To pack off the fish when catching with a spear, it is necessary to place an extension below the spear with the appropriate pack-off cups facing down (Figure 7-6). Frequently these are protected by a steel guide which helps the pack-off cups to enter the pipe without damage.

A stop sub is frequently run above the spear to space it in the fish properly. It is desirable to place the grapple far enough in the fish to secure a good grip, but if it is too far below the top of the pipe, it may swell the pipe if a large force is applied. Releasing the spear would then become a problem. Ordinarily the stop sub is placed one to two feet above the catching surface of the spear. However, extensions may be used to place it lower in the fish if the top is splintered or swelled.

There are other designs of spears with variations from the preceding description, but most are built on the principle of the tapered wedge. There are "J" releases, as well as "automatic bottoms" or split nuts and cams that are used to set and release the slips or grapple.

Chapter 8
Jarring Stuck Pipe or Tools

Jars are impact tools used to strike heavy blows either up or down upon a fish that is stuck. Jars have been used in drilling for ages as the cable tool drillers used link jars for both fishing and drilling. Today, jars fall into two categories as to use: *fishing jars* and *drilling jars*. While they are both basically designed in accordance with the same principle, they are usually built quite differently. This will be explained further in the discussion. Each of these jars, classified according to use, can be further separated according to the basic principle of operation; either *hydraulic* or *mechanical*.

Most jarring strings (Figure 8-1) for fishing consist of an oil jar (sometimes called *hydraulic jar*) and a bumper jar (also called a *bumper sub*) along with the necessary drill collars for weight. In addition to these, an accelerator (also called an intensifier or booster) may also be added to the string.

The oil jar is designed to strike a blow upward only, while the bumper jar is designed to strike a blow downward on the fish. The accelerator (intensifier or booster) may be included in the jarring string to provide additional stored energy which helps to speed up the travel of the drill collars when released by the oil jars. It also provides free travel which compensates for the travel of the oil jar mandrel. This travel compensation prevents the work string from being pushed up the hole, which absorbs the energy of the impact through friction.

Bumper Jar

The bumper jar (Figure 8-2) is a mechanical slip joint. The jars are manufactured as simple models in which either the mandrels are exposed when open or the mandrel splines are enclosed and lubricated. The bumper jar is almost exclusively used as a down impact tool. The weight

36 Oilwell Fishing Operations

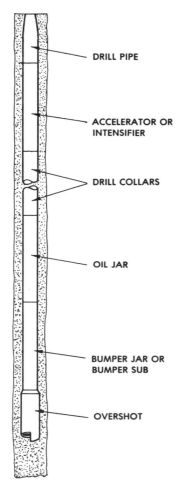

Figure 8-1. Typical jarring string showing sequence of tools.

of the drill collars is released suddenly, causing a heavy impact as the bumper jar closes. In addition to delivering impact blows to the fish, bumper jars are used above catching-type tools such as overshots and spears. If the tapered grapples or slips become stuck on the mandrels or in the bowls, they may be jarred down off the tapers by the bumper jars. This is necessary in order to release the tool from the fish.

Fishing tool operators will frequently use bumper jars in a string of fishing or cutting tools so that a constant weight may be applied to a tool such as a cutter. By operating within the stroke of the bumper jar, only the weight below that point is applied to the tools below. An example would be a cutter in a deviated hole. The weight run below the bumper jar could be exerted on the knives, but excessive weight from the work string could be avoided.

Oil Jar

The oil jar (Figure 8-3) consists of a mandrel and piston operating within a hydraulic cylinder. When the oil jar is in the closed position, the piston is in the down position in the cylinder where it provides a very tight fit and restricts the movement of the piston within the cylinder. The piston is fitted with a unique set of packing which slows the passage of oil from the upper chamber to the lower chamber of the cylinder when the mandrel is pulled by picking up on the work string at the surface. About half way through the stroke, the piston reaches an enlarged section of the cylinder and is no longer restricted so the piston moves up very quickly and strikes the mandrel body. The intensity of this impact can be varied by the amount of strain taken on the work string. This variable impact is the main advantage of the oil jar over the mechanical jar for fishing.

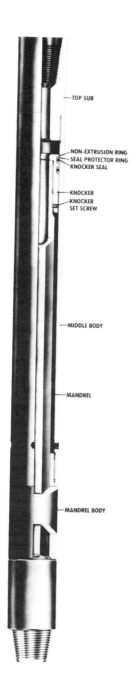
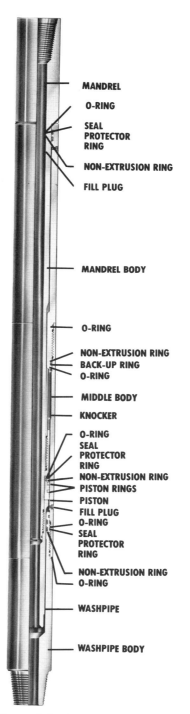

Figure 8-2. Fishing bumper jar shown in closed position. (Courtesy of Bowen Tools.)

Figure 8-3. Fishing oil jar shown in closed position. (Courtesy of Bowen Tools.)

Most oil jars are completely effective to 350°F but they may be secured with a special heat-resistant oil which will sustain higher temperatures.

Newer models of oil jars are designed with check or bypass valves which allow the fluid to quickly transfer to the chamber above the piston when cocking or reloading. However, there are many oil jars in the field that do not have this feature. In the older models, the fluid must transfer through the gaps in seals and rings. If weight is rapidly applied to close them, the fluid will pass around the seals and destroy them shortening the life of the jars. Caution should be used to slack off weight slowly when reloading the jar to prevent this damage.

Current types of oil jars also incorporate a floating piston which effectively transfers the pressure of the hydrostatic head to the jar fluid.

Oil jars are very effective in freeing stuck fish as the energy stored in the stretched drill pipe or tubing is converted to an impact force. This can easily be varied according to the pull exerted on the work string.

Jar Intensifier or Accelerator

The intensifier or accelerator (also called a booster jar) (Figure 8-4) is an accessory run in the jarring string. When run above the drill collars, the impact delivered to the fish is increased and most of the shock is relieved from the work string and rig.

The tool is essentially a fluid spring. It is a cylindrical tool filled with a compressible fluid, usually an inert gas or silicone. When the work string is pulled, a piston in the cylinder compresses the fluid and stores up energy. When the oil jar trips, this energy is released and speeds the movement of the drill collars up the hole thereby imparting a heavier blow.

The other function of the accelerator is to relieve the work string of the majority of rebound which is damaging to tools and tool joints. This is accomplished by the free travel available when the accelerator is pulled open. The free travel in the accelerator compensates for the free stroke of the oil jar. Ordinarily, without an accelerator in the string, the work string is stretched and when the oil jar trips, the pipe is released to move up the hole where much of the stored energy is absorbed in friction in the wellbore. This is made apparent by the movement at the surface, causing the elevator, traveling block, and even the derrick to shake. This movement does not occur with an accelerator in the string due to its compensation of the free travel of the oil jar. One worthwhile advantage of running an accelerator is preventing this sudden compressive force from being exerted on the work string. Since the impact is increased due to the

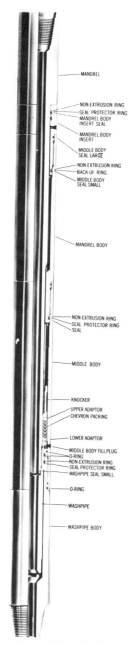

Figure 8-4. Jar intensifier shown in closed position. (Courtesy of Bowen Tools.)

higher speed with which the drill collars move up to strike a blow, less weight or mass is required to impart the desirable impact. Manufacturers furnish recommended weights of drill collars to be run with each oil jar. It is important when running an accelerator not to exceed the recommended weight as the efficiency is increased so much that tools or the fish may be parted without the desired movement up the hole.

Jarring Strings

Complete jarring strings (Figure 8-1) consist of (from bottom) the appropriate catching tool or screw-in sub, the bumper jar or sub, the oil jar, desired drill collars, accelerator jar, and the work string. Each tool in the string performs a specific function, and it is necessary that they be run in that specific sequence. The overshot or spear on bottom is to catch or engage the fish. The bumper jar is for jarring down, either to jar the fish or to help release a catching tool. The oil jar imparts the blow up, the drill collars furnish the weight necessary for a good impact, and the accelerator speeds up the jarring movement and compensates for the travel in the oil jar, saving the pipe from the compressive stresses.

The weight of drill collars run with jarring strings varies according to the sizes of the jars used but is also influenced by the depth of the fish, the fluid in the well, the strength of the work string, and the amount of fish stuck in the wellbore. Manufacturers of oil jars and booster jars make recommendations for a range of weights to be run with their tools. If this information is not available, a rule of thumb that has been used, particularly in cased holes, is to run the number of drill collars corresponding to the diameter of the jar in inches. Therefore a 4 3/4-in. O.D. jar with 3 1/2-in. connections would indicate four or five 4 3/4 in. O.D. drill collars which weigh approximately 1,500 lb each. This compares favorably with the manufacturer's recommendation of 5,600 lb to 7,500 lb of drill collar weight for this jar. The manufacturer recommends 11,800 lb to 16,000 lb of drill collar weight with a 6¼-in. O.D. jar with 4½-in. IF connections. Drill collars of 6¼-in. O.D. weigh approximately 2,700 lb each, so six drill collars of this size are exceeding the recommendation by a slight amount. This rule of thumb is good only through sizes with 3½-in. connections.

When accelerator jars are run, it is important not to run excessive weights of drill collars as they tend to overload the accelerator and hinder rather than help the action. Do not run drill collars or heavyweight pipe above the accelerator. This procedure causes the string to elongate when the jars trip, and may cause an impact in the accelerator which is damaging to the tool. If it is necessary to include drill collars or heavyweight pipe in the string for length, they may be placed up the hole where they would not affect the jarring operation.

up to a predetermined amount above its weight and the brake set while the oil jar bleeds off and the blow is delivered. This jarring weight may be varied to any amount within the strength of the tools and pipe run. Ordinarily jarring is started at low impact values and is gradually increased as necessary. This variable impact is the big advantage of hydraulic or oil jars over mechanical jars. When running an overshot or other catching tool, it is desirable to start jarring at a low impact and increase as necessary. This tends to set the grapple and allow it to bite into the fish. If a heavy blow is struck first, it can cause the grapple to strip off the fish, leaving the grapple dull and unable to catch again.

In jarring up, the bumper jar has no function and merely acts as an extension or slip joint. When jarring down is desirable, the oil jars should be closed and the stroke of the bumper jar used for the downward impact. This precaution is necessary to prevent jarring down on the oil jar packing, which would cushion the downward impact and perhaps damage the oil jar packing.

Jars are always redressed after each use, even if they did not strike a blow. They are disassembled, and inspected, and new seals and oil are installed. Then they are tested on a pull rack for resistance to pull according to the size.

Two oil jars are never run together, as they do not trip at the same time and one would impact the packing and seals of the other. This would be destructive to the packing and counter productive to good jarring action.

Jars should always be replaced when making a trip for some purpose other than to change them out. There is no way to determine the remaining useful life of a jar that has been operated, therefore the replacement is the best insurance.

Considerable research has been done, and is continuing, on the proper location and configuration of jarring strings. Manufacturers of jars have always recommended drill collar weight to run above the oil jar for acceptable impact. In 1979, Skeem, Friedman, and Walker published their paper "Drillstring Dynamics During Jar Operation." This basic study of stress wave mechanics opened the door for further studies and modeling of the various situations encountered in fishing. It has become apparent that the impact duration or impulse may be as important as the peak force or impact magnitude.

In the past it was thought that the addition of drill collars or heavyweight drill pipe above the oil jar created an increase in impact, but it was not always possible to move this mass fast enough to be effective. The impulse or duration of the impact is also important, particularly for long sections of stuck fish such as differentially stuck drill collars.

A very definite opinion has been formed from these studies: drill collar weights recommended by the manufacturers should not be exceeded, and efforts should be made to keep the weights low in the range of such recommendations.

Some manufacturers are now able to analyze specific jobs by computer, and make recommendations for the design of jarring strings.

Surface Jars

On some occasions, the drill string becomes stuck near the surface, primarily in key seats, or hung in the bottom of the surface pipe. To free this pipe, it is desirable to strike a heavy blow down since jarring upward would only cause it to become stuck further. Early drillers fashioned a "driving joint" consisting of an old kelly or joint of pipe and a sleeve or large pipe outside and sliding on the inner body. Two flanges, one on each member, were used as the striking faces. This was made up in the string at the surface, and then the outer body was picked up with the catline and dropped. This imparted a good impact against the flange made up in the string and frequently freed it, saving an expensive fishing job.

The driving joint has been replaced in most operations with a surface bumper jar (Figure 8-5), which also imparts a heavy down blow. By adjusting the friction slip in the jar, the impact may be increased or decreased as needed. The jar is made up in the string at the surface and the friction slip and control ring adjusted to the estimated desired tripping pull. When a straight upward pull is exerted on the jar, the friction slip rubs the enclosed friction mandrel and arrests upward movement while the drill pipe is being stretched. When the upward pull reaches the preset tripping tonnage, the friction mandrel is pulled through the friction slip.

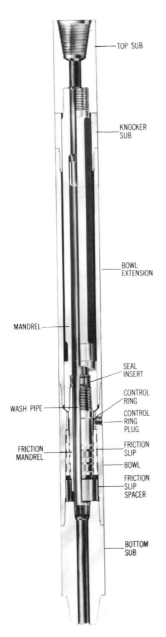

Figure 8-5. Mechanical surface jar. (Courtesy of Bowen Tools.)

The resulting downward surge of the pipe in returning to its normal length causes a sudden separation of the main mandrel and bowl assemblies which are free to move apart for the length of its 48-in. stroke and drive the weight of the free pipe against the stuck point.

As in all jarring operations, light blows should be used at first. If light jarring is unsuccessful, the tool may be adjusted for heavier impacts. The tripping tonnage should never be set higher than the weight of the free pipe between the jar and the stuck point. If the jar is set higher than this weight, it becomes necessary to pull on the pipe at the stuck point which will usually cause it to stick more.

Occasionally, fishing tools that operate on tapers may become frozen. The wickers on grappling tools may bite into the fish sufficiently that the tools cannot be released in the normal manner. A bumper jar with a downward impact is very effective in freeing the grapple so that the tool may be released. The surface bumper jar is also used for this purpose.

Drilling Jars

Under some drilling conditions, it is economical to run jars in the drill string so that they are readily available if the string becomes stuck. As stated earlier, these jars can be divided into two types, according to design.

The hydraulic or oil jar (Figure 8-6) is essentially the same design as the fishing oil jar except that it is made much heavier and stronger in order to sustain several hundred hours of drilling. The down jar is essentially a mechanical detent design using friction slips and mandrels. A compari-

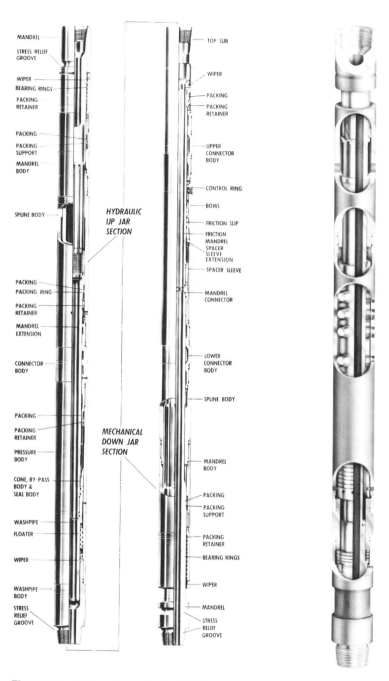

Figure 8-6. Hydraulic-mechanical drilling jar. (Courtesy of Bowen Tools.)

Figure 8-7. Mechanical drilling jar. (Courtesy of Dailey Oil Tools, Inc.)

son of drilling jars with the fishing jars shows the basic differences in size:

	Length	Stroke	Approx. Wt.
4½ FH Drilling Jar	36 ft approx.	Oil—13½ in. Mech.—7 in.	3,800 lb
4½ FH Fishing Jar	8 ft to 11 ft ea.	Oil—4⁵/₁₆ in. Bumper—18 in.	1,100 lb

In the preceding comparison, both up and down jarring tools are included.

The mechanical drilling jar (Figure 8-7) is manufactured in several styles and uses a pull on the drill string in some manner to trip. One model uses the torsion bar principle. As pull or weight is applied, rollers force a slotted sleeve to rotate, allowing the mandrel to become free in the stroke. The torque is preset by a series of springs, and it can be varied slightly by applying torque to the drill pipe at the rotary. Right-hand torque increases the pull necessary for the jar to trip while left-hand torque decreases the necessary weight or pull.

Another model of mechanical drilling jars uses slots and lugs in the mandrel and body to trip and to set the jar. During ordinary drilling, the lugs are engaged in the slots. If sticking occurs, a pull is exerted on the drill pipe and then torque is applied. The lugs slip out of the slots, and jarring occurs.

Drilling jars should ordinarily be run in tension above the neutral point of the string. If placed at the transition zone, they would be subjected to unusual flexing of the tool, causing premature failure. Run above the majority of drill collars, the jars are readily available if the bit or collars stick. Several drill collars or heavy pipe may be run above the jars to increase the impact due to additional mass. Each manufacturer furnishes instructions with the particular design featured.

Impact forces of jars are expressed in terms of pounds jarring. This is purely theoretical and is an expression of the amount pulled on the work string above its normal weight. True impact will vary with many variable conditions. Mud weight, friction in the hole, drill collar mass or weight, and the stroke of the jars will affect the true impact. Certain theoretical calculations have been made and are used primarily to prevent jarring with too much pull and/or weight. It is desirable to move the fish and not to hit it so hard that it parts.

Chapter 9
Washover Operations

Washover Pipe

Washover pipe, or washpipe as it is commonly called, is large pipe used to drill out, wash out, and to circulate cement, fill, formation, or other debris that is causing the fish to be stuck. Washpipe of the proper size to be safe in the specific operation must be selected. It must be of sufficient inside diameter to go over the fish with clearance for circulation, of an outside diameter small enough to rotate in the hole or casing, and with an annular clearance sufficient for circulation and prevention of over-torquing and thus the sticking of the washpipe.

Note Table 9-1, which lists sizes of washpipe and the hole size or casing O.D. in which these sizes are normally used, as well as the maximum size of fish that it is safe to cover with that diameter of pipe.

A washpipe string is made up of a top bushing or safety joint, the number of joints of pipe that it has been determined to run, and a rotary shoe on bottom designed for the particular material to be cut (i.e., formation, fill, cement or steel).

The top bushing is merely a substitute from the washpipe threads to a tool joint appropriate to fit the running string. Safety joints (Figure 9-1) are sometimes run in place of the top bushing, but bushings have become more popular in the last few years. There are several reasons for this:

- Safety joints are not always reliable.
- String shots have become more popular, and they are usually reliable in that a back-off can be made where intended.
- There are now back-off connectors and washpipe spears run with the washpipe string which require some left hand torquing—this would not be possible with a safety joint in the string.

The washpipe itself is usually heavy wall N-80 grade casing cut into Range 2 lengths for ease of handling and equipped with special threads

Oilwell Fishing Operations

Table 9-1
Typical Washover Pipe Specifications

Size And Connection (in.)	Weight Plain End (lb/ft)	Inside Dia. (in.)	Wall Thick (in.)	Upset Dia. (in.)	Drift Dia. (in.)	Make-Up Torque (Foot-Pounds) Rec.	Make-Up Torque (Foot-Pounds) Max.	Joint Tensile Yield Strength (lb) +	Joint Efficiency %	Fish. O.D. Washover Size (in.) Rec.	Fish. O.D. Washover Size (in.) Max.
3½ WP	8.81	2.992	.254		2.867	850	3,400	97,400	47	$2^{11}/_{16}$	$2^{7}/_{8}$
3¾ WP	9.54	3.238	.256		3.113	1,015	4,060	108,700	45	3	$3^{1}/_{8}$
4 WP	12.93	3.340	.330		3.215	1,370	5,600	153,670	51	$3^{1}/_{16}$	$3^{1}/_{4}$
4⅜ WP	13.58	3.750	.313		3.625	1,660	6,650	158,700	50	$3^{1}/_{2}$	$3^{5}/_{8}$
4½ WP	13.04	3.920	.290		3.795	1,460	5,860	160,800	52	$3^{5}/_{8}$	$3^{3}/_{4}$
4½ WP	14.98	3.826	.337		3.701	1,800	7,220	181,200	51	$3^{1}/_{2}$	$3^{11}/_{16}$
5 WP	14.87	4.408	.296		4.283	1,870	7,500	184,600	53	4	$4^{1}/_{8}$
5 X-Line	15.00	4.408	.296	5.360	4.151	2,700	3,000	187,000	53	4	$4^{1}/_{4}$
5 X-Line	17.93	4.276	.296	5.360	4.151	2,700	3,000	259,000	61	4	$4^{1}/_{8}$
5 WP	17.93	4.276	.362		4.151	2,460	9,850	218,532	52	4	$4^{1}/_{8}$
5½ WP	16.87	4.892	.304		4.767	2,370	9,480	209,700	52	$4^{5}/_{8}$	$4^{3}/_{4}$
5½ X-Line	17.00	4.892	.304	5.860	4.653	2,700	3,000	220,000	55	$4^{5}/_{8}$	$4^{3}/_{4}$
5½ WP	20.00	4.778	.361		4.653	2,970	11,900	237,200	51	$4^{1}/_{2}$	$4^{5}/_{8}$
5¾ WP	18.12	5.125	.313		5.000	2,700	10,800	222,800	52	$4^{7}/_{8}$	5
5¾ PSI	21.53	5.000	.375		4.875	3,400	13,580	246,500	49	$4^{3}/_{4}$	$4^{7}/_{8}$
6 WP	19.64	5.352	.324		5.227	3,170	12,700	238,800	52	$5^{1}/_{8}$	$5^{1}/_{4}$
6⅜ WP	24.03	5.625	.375		5.580	4,250	17,000	288,300	52	$5^{3}/_{8}$	$5^{1}/_{2}$
6⅝ WP	23.58	5.921	.352		5.796	4,400	17,590	251,600	45	$5^{5}/_{8}$	$5^{3}/_{4}$
7 WP	25.66	6.276	.362		6.151	4,970	19,880	315,200	52	6	$6^{1}/_{8}$
7 X-Line	26.00	6.276	.362	7.390	6.151	3,200	3,500	351,000	58	6	$6^{1}/_{8}$

Washover Operations

Table 9-1 (continued)
Typical Washover Pipe Specifications

Size And Connection (in.)	Weight Plain End (lb/ft)	Inside Dia. (in.)	Wall Thick (in.)	Upset Dia. (in.)	Drift Dia. (in.)	Make-Up Torque (Foot-Pounds)		Joint Tensile Yield Strength (lb) +	Joint Efficiency %	Fish. O.D. Washover Size (in.)	
						Rec.	Max.			Rec.	Max.
7³/₈ WP	28.04	6.625	.375		6.500	5,725	22,900	341,300	52	6³/₈	6¹/₂
7⁵/₈ WP	29.03	6.875	.375		6.750	6,120	24,500	355,000	52	6⁵/₈	6³/₄
7⁵/₈ WP	33.04	6.765	.430		6.640	7,520	30,100	398,900	51	6¹/₂	6⁵/₈
7⁵/₈ X-Line	29.70	6.875	.375	8.010	6.750	3,700	4,000	385,000	56	6⁹/₁₆	6¹¹/₁₆
8¹/₈ WP	31.04	7.375	.375		7.250	6,820	27,300	373,900	51	7¹/₈	7¹/₄
8¹/₈ WP	35.96	7.250	.437		7.125	8,370	33,500	398,500	51	7	7¹/₈
8³/₈ WP	33.95	7.578	.399		7.453	7,500	30,000	404,900	47	7¹/₄	7³/₈
8⁵/₈ WP	36.00	7.921	.352		7.796	6,700	27,100	335,300	51	7¹/₂	7¹¹/₁₆
8⁵/₈ X-Line	36.00	7.781	.400	9.120	7.700	4,200	4,500	567,000	41	7⁹/₁₆	7⁵/₈
8⁵/₈ WP	39.29	7.725	.450		7.600	9,950	39,800	475,200	65	7¹/₂	7¹¹/₁₆
9 WP	38.92	8.150	.425		7.994	9,920	39,700	458,400	51	7⁷/₈	8
9⁵/₈ X-Line	43.50	8.665	.560	10.100	8.599	4,700	5,000	701,000	50	8³/₈	8¹/₂
9⁵/₈ WP	43.5	8.681	.472		8.525	13,000	52,100	543,800	64	8¹/₄	8¹/₂
10³/₄ WP	51.00	9.850	.450		9.694	15,600	62,400	595,600	50	9³/₈	9⁵/₈
11³/₄ WP	58.81	10.772	.489		10.616	19,250	77,000	684,300	51	10	10¹/₂
13³/₈ WP	72.00	12.415	.480	13.750	12.259	36,500	146,100	1,147,600	49	11¹/₂	12
16 WP	75.00	15.010	.495		14.823	62,580	250,300	1,791,600	52	14¹/₄	14³/₄
16 WP	109	14.688	.656		14.50	62,580	250,300	1,800,000	52	13¹/₂	14¹/₄

Figure 9-1. Washover safety joint. (Courtesy of Texas Iron Works.)

Figure 9-2. Special washpipe thread. (Courtesy of Hydril Company.)

with good characteristics for torquing and strength. (See Figure 9-2.) When a washover operation is conducted, it is actually a drilling procedure, and therefore the pipe is subjected to high torque. Usually shoulders, such as those used on tool joints, are included in the design of the washpipe thread so that it will not fail with the high torque required. Maximum cross section of the threads is also designed into the special joint for high thread efficiency. This special thread design is very necessary, since ordinary tapered casing threads would continue to make up during the rotating and a failure would occur.

Washpipe is usually flush joint both inside and outside for maximum clearance. This type is also run inside casing for work-over operations. In areas where differential sticking is a problem in open hole work, pipe with either an external upset or a coupling is used. The X-line joint (Figure 9-3) is popular as an external upset connection and the Brown Oil Tool joint is an example of the "collared" pipe.

Washover Operations 49

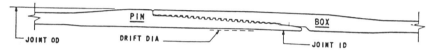

Figure 9-3. Extreme-line casing thread.

Rotary Shoes

The rotary shoe (Figure 9-4) run on the bottom of the washpipe string should be designed for the particular job. Tooth-type shoes are usually used if cuttings, fill, formation, or cement is to be cut. The teeth are shaped with a straight leading edge, and all the surfaces of the teeth are dressed with a wear material, usually tube borium, to prevent excessive wear and erosion from the fluids circulation.

If steel, such as the tool joints, tube, or junk, must be cut by the rotary shoe, it is dressed with tungsten carbide in a configuration that is appropriate for the particular job. Care should be exercised in designing the shoe, since it is necessary to have sufficient circulation to keep the carbide cool as well as wash away the cuttings. If the job is inside casing, no cutting carbide should be allowed to remain on the outside as this will damage the casing. In some cases, smooth brass is applied on the outside diameter of the shoe to provide a bushing, which reduces the friction and prevents damages to the casing. Tungsten carbide is applied to the bottom of the shoe and if possible to the inside diameter. Where it is possible to apply the carbide with a small shoulder inside the shoe, the chances of retrieving some or all of the fish inside the pipe are improved. This would save a trip with another tool to recover what has been washed over.

The length of the washpipe string is most important. Realizing that the washpipe is large, stiff, and smooth, length becomes extremely important in preventing sticking. There is no rule or gauge for determining the maximum length, but a judgment must be made based on careful consideration of the hole conditions.

Two actual jobs which demonstrate the extremes in lengths of washpipe strings are described in the following.

On the first job, the drill pipe was stuck from a depth of 330 ft (the bottom of the surface pipe) to 8,487 ft (the depth of the bit). Obviously the cause of sticking was a poor mud system, therefore the job consisted of circulating out this mud and replacing it with a suitable mud system, conditioning the hole as progress was made. On the last washover, 1,218 ft of washpipe was run. This is unusual, but nevertheless, under the circumstances of this particular job, the decision was correct and the job was completed in a satisfactory manner.

50   Oilwell Fishing Operations

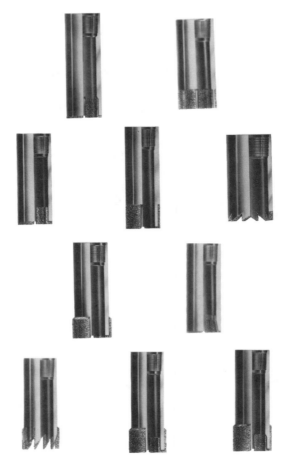

Figure 9-4. Rotary shoes.

In the foregoing situation there was a dollar value of equipment as well as the cost of drilling the hole, the cost of the surface pipe, and the cost of cementing it to be recovered. To be economically successful, the washover operation cannot cost more than the cost of replacing the hole and equipment lost in it.

Another actual job consisted of washing over 47 joints of 3½-in. drill pipe cemented in a 7-in. liner at approximately 14,000 ft and in a 36° hole. This job was completed satisfactorily but 5 joints or approximately 150 ft of washpipe were all that could be run at one time. When six joints were run, it became stuck and added expense in milling and fishing the washpipe. Another problem arose when the washpipe was stopped from rotation or reciprocation in order to run a wash-out tool. The washpipe became stuck due to the cement cuttings settling out. Deviated holes or accidentally crooked holes limit the length of the safe washpipe string.

Probability percentages must be applied to the cost formulas. Certainly every job is not successful. Success is expected, but what is the probability that it will occur? Many operating companies apply probability when making such decisions and setting up parameters for the job. Probability percentages should be determined from specific experience in a number of similar jobs. In field drilling and in workover programs within the same field, records will indicate what problems have developed and the frequency that they have occurred. It is only from actual results that reliable probability factors can be developed.

When the entire length of fish cannot be covered in one washover, it is necessary to part the string that has been freed from that which remains in the hole. This can be accomplished by one of several methods:

1. An overshot can be run after the washpipe has been removed and left-hand torque applied and the fish backed off with a string-shot. (See Chapter 6.)
2. An external or outside cutter can be run on the washpipe instead of the rotary shoe and the fish cut off above the lowest point that has been freed.
3. A washpipe spear may be run in the washpipe string during the washing over, and the spear can then be used to apply the left-hand torque for the string-shot back-off.
4. A back-off connector may be run in the top joint of washpipe and engaged in the top of the fish when the washover is completed. Through this connector, left-hand torque can be applied and the string-shot back-off made.

External Cutters

The outside or external cutter (see Figure 9-5) is usually slightly larger in outside diameter than the washpipe, and it is dressed to catch the type of tool joints or couplings that are on the fish.

Pipe with couplings uses a catcher assembly with spring fingers that catch below the coupling.

Pipe with couplings but with upset joints can be caught with dog-type or pawl-type catchers (Figure 9-6) made with slip surfaces cut on the end where they will engage the upsets. Flush joint pipe requires a hydraulically actuated catcher. Pump pressure against the sleeve restriction in the annular space actuates the knives (see Figure 9-7). Note in Figure 9-5, as the washpipe is moved upward, the finger catcher assembly near the top

Figure 9-5. External cutter with spring dog catcher for coupled pipe. (Courtesy of Bowen Tools.)

of the cutter engages the pipe under the coupling and in turn this sleeve moves down the barrel, transmitting force through the spring to the sleeve which feeds in the knives. As this assembly rotates, the knives cut the pipe in two. Coil springs as shown in the illustration are used in practically all cutting tools as they tend to absorb heavy shocks which prevents breaking the knives.

Washpipe Spears

The washpipe spear is used primarily to prevent dropping a fish that is stuck off bottom when washing over. It is very versatile, however and can be used effectively to pick up a fish on the same trip as the washover and to back off a fish when it is partially washed over, thus saving a trip. Ordinarily, when a fish is stuck off bottom and it is washed over, it falls to bottom and may corkscrew the pipe, damage the bit by knocking off cones or even shanks, and damage the filter cake in the wellbore. Catching the fish when it is freed and preventing its fall may be very important and cost effective.

The spear (Figure 9-8) is made up of two major assemblies: the mandrel with the slip mechanism and the control cage with the friction blocks, restriction rings, and latch. The spear is placed in the washpipe string, usually in the bottom joint, but it may be run anywhere in the washpipe string that is desired. It is anchored in place by turning the bottom sub to the left (or counterclockwise), which moves the tapered slip cone under the slips, extending them and anchoring the spear in the washpipe. Below the spear, an unlatching-type safety joint is run. When the washpipe has been worked over the top of the fish and the spear engaged in the fish, the connection is made up by the usual right-hand rotation.

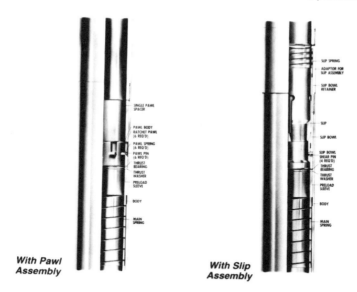

Figure 9-6. External cutters for upset pipe. (Courtesy of Bowen Tools.)

By continuing the rotation with an upward pull, the slip cone is entirely retracted and the slips will not drag on the washpipe. The spear is now firmly connected to the fish and is not engaged in the washpipe except through the friction blocks on the control cage. The usual pump pressure in washing over is applied against the surface of the restriction rings which holds the cage down on the mandrel. If the fish is cut loose and starts to fall, the friction blocks hold the cage firmly in the washpipe and the mandrel moves down. Without the cage holding the slips in a retracted position, a spring below the slips moves them up, engaging the washpipe and stopping any further downward movement of the mandrel and the fish.

The use of the washpipe spear will prevent a stripping job when the fish is recovered. At the surface, when the fish is reached inside the washpipe, drill pipe may be picked up, made up handily in the top of the spear, and the spear manually latched off or disengaged. The spear with the fish is then lowered to the bottom of the washpipe and set in the pipe there, and the drill pipe removed. The fish is now hanging out of the bottom of the washpipe and a time-consuming stripping job has been prevented.

If the fish cannot be freed in one washover, the spear may be actuated by shutting down the pumps and picking up on the washpipe. Torque can now be applied through the washpipe and the spear so that a string-shot

54 Oilwell Fishing Operations

Figure 9-7. External cutter with hydraulically actuated catcher. (Courtesy of Bowen Tools.)

back-off may be made. When the pipe has been parted, the freed portion is brought to the surface by the spear and the washpipe.

When a fish is leaning over in a washed out section of the wellbore, and it is not possible to go over it with the usual washpipe assembly, a joint of pipe, slightly bent, can be run below the washpipe spear. With this flexible pipe hanging below the washpipe, it becomes easier to stab into the fish.

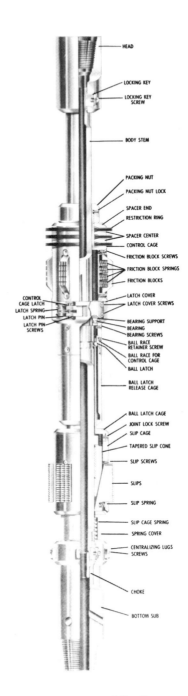

Figure 9-8. Washpipe or drill collar spear. (Courtesy of Bowen Tools.)

Unlatching Joint

The spear is always run with an unlatching or "J" type safety joint immediately below it. The unlatching joint is held firmly in place with two light metal straps which prevent it from accidentally unlatching while going in the hole. After the spear and safety joint are made up in the fish, straight pick-up pulls the straps apart and the safety joint is then operational. It is usually dressed so that right-hand torque can be transmitted through it, a straight pull exerted on it, and unlatching accomplished by slight left-hand torque while picking up. By running the safety joint below the spear, the washpipe spear can be brought out of the hole with the washpipe at any time a trip is made to exchange rotary shoes or for any other reason.

Back-Off Connector

When washing over and retrieving long strings of pipe that are resting on bottom, a back-off connector may be used to advantage to reduce the number of trips with the work string. This assembly is essentially a J-type unlatching joint made up inside the top joint of washpipe, and screwed into a box connection on the lower side of the washpipe safety joint. The connector is subbed to the proper thread to screw into the top of the fish. When contact is made with the top of the fish, after washing down to it, a connection may be made with the back-off tool. After the connection is made, circulation may be established (unless the fish is plugged), the J-type unlatching joint may be parted, and pipe may be substituted for the kelly joint. With the left-hand torque transmitted through the back-off connector, a string shot back-off may be made and the fish retrieved on the same trip.

Hydraulic Clean-Out Tools

Occasionally during the recovery of stuck drill pipe or tubing, the inside of the pipe becomes bridged over. This prevents the lowering of string shots or electric wireline cutting tools to the desired depth. Spudding with the wireline and an assembly on bottom to cut through the bridge is sometimes successful if the bridge is not too long or compacted. When the bridge cannot be removed by spudding, the next choice would ordinarily be to use a hydraulic clean-out tool.

Note in Figure 9-9 that the tool consists of a jet-type shoe, lengths of the clean-out tubing, usually $1^3/_8$-in. O.D., a top sub which may include a stop ring and entry circulating ports, and a connection to the sinker bars, rope socket, or the free-point indicator, if it is run at the same time.

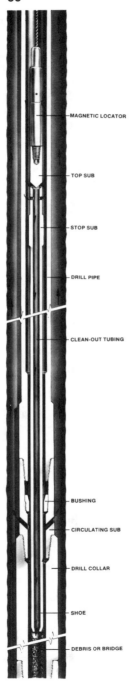

After a back-off or cut is made in the free pipe, the pipe is pulled and a circulating sub of the correct size and thread is installed on the bottom of the pipe string. It is then run back in the hole and screwed into the fish. If the pipe is cut instead of backed off, an overshot without a pack-off is used to tie back to the fish.

Up to 300 ft of the small tubing may be run, and when it reaches the bridged area, the pumps are started and a jetting action washes away the plugging material as the clean-out tool is lowered on the wireline. When this has been accomplished, the clean-out tool may be removed from the well and normal pipe recovery methods used again.

Figure 9-9. Hydraulic clean-out tool. (Courtesy of N. L. McCullough.)

Chapter 10
Loose Junk Fishing

If at all possible the first step in fishing any loose junk from the hole is to identify what it is. This may be readily determined if something has been left in the hole on a trip or has been dropped into the hole accidentally. If the type and configuration of the junk are not known, an impression block should be considered. See Figure 21-4. It will also help to visualize how the fish may be retrieved if another part or tool exactly the same as the fish is placed in a casing nipple of an appropriate size to simulate the hole. The proposed fishing tools may then be tried at the surface, and those found inappropriate may be ruled out. It is much cheaper to try the proposed tool on the surface than to make a trip and retrieve nothing.

The usual tools for retrieving loose junk are magnets, various types of junk baskets, hydrostatic bailers or a tool that might be fashioned for the particular circumstance.

Magnets

Fishing magnets are either permanent magnets fitted into a body with circulating ports or electromagnets which are run on a conductor line.

Permanent magnets (Figure 10-1) have circulating ports around the outer edge so that fill and cuttings can be washed away and contact made with the fish. Ordinarily the magnetic core is fitted with a brass sleeve between it and the outer body so that all of the magnetic field is contained and there is no drag on the pipe or casing. Permanent magnets have the advantage of the circulation washing away any fill so that the junk is exposed. Ordinarily, by rotation, one can detect when contact is made with the fish. The operator should then thoroughly circulate the hole, shut the pump off, and retrieve the fish. When pulling the work string, it should not be rotated as there is a chance of slinging the fish off.

58 Oilwell Fishing Operations

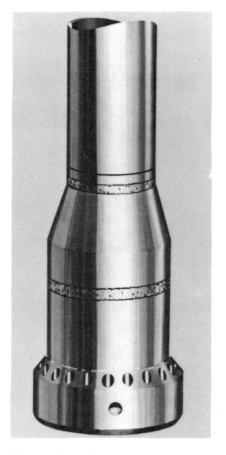

Figure 10-1. Permanent fishing magnet. (Courtesy of Bowen Tools.)

Figure 10-2. Fishing electromagnet. (Courtesy of Schlumberger.)

Magnets are usually furnished with a cut-lip guide, a mill tooth guide (which is the most popular), and a flush guide, which resembles a thread protector. The guide that extends down below the magnet is extremely helpful in retaining the fish and protecting it from being dragged off inside the casing.

Electromagnets (Figure 10-2) are run on a conductor line and charged only when the bottom of the well is reached. They have the advantage of quick trips in and out of the hole plus the added lifting power of an electromagnet. However, if the fish is covered with fill or debris, it cannot be reached, since there is no way to circulate the tool down.

Magnets will only pick up ferrous metal. Other methods should be used to recover brass, aluminum, carbide, and stainless steel.

Junk Baskets

Core-Type

This was the old stand-by for years for fishing bit cones and similar junk from an open hole. It consists of the top sub or bushing, a bowl, a shoe, and usually two sets of finger-type catchers (Figure 10-3). This tool is still used quite often, and it is made to circulate out the fill and to cut a core in the formation. The two sets of catchers, one dressed with short fingers, help to break the core off and retrieve it. Any junk that is in the bottom of the hole is retrieved on top of the core.

With all catcher-type junk baskets, the catcher must rotate freely in the bowl or shoe. As the tool is run down over junk, rotating and circulating, the catcher snags on the junk and remains stationary as the bowl and shoe rotate around it. If the catcher is fouled with trash, excessive paint or corrosion, or other foreign material, it will not rotate and the fingers will be broken off, resulting in additional junk in the well bore.

Reverse Circulation

In many workover operations, we "reverse circulate" the fluid by pumping it down the annulus and returning it through the work string. This makes it possible to circulate out larger and heavier particles than when pumping the "long way," or pumping down the work string with returns through the annulus. In many operations, the workover fluid need not be nearly as viscous if it can be reverse circulated. However, in open holes, it is seldom possible to reverse circulate due to pumping into the formation. Nevertheless, the reversing action is extremely helpful in kicking into the barrel and catcher junk that might other-

Figure 10-3. Core-type junk basket. (Courtesy of Bowen Tools.)

60 Oilwell Fishing Operations

wise be held away from the catcher by the fluid flow. With this in mind, in recent years, two different designs of reverse circulation junk baskets have been introduced.

The first design incorporates jets or venturis that are opened when a ball is dropped down the work string (Figure 10-4). The space behind a jet has a reduced pressure, therefore, as the jet forces fluid from the bowl into the annulus, the interior of the bowl is at a lower pressure causing wellbore fluid to enter through the catcher.

The other design incorporates an inner barrel with the fluid flow between the outer and inner barrels when a ball is dropped and closes the center flow through the seat (Figure 10-5). In this design, after washing the wellbore sufficiently to remove all fill, the ball is circulated down, and when it seats the fluid flow is diverted between the two barrels and flow is actuated through the upper ports into the annulus from the inner barrel. Reverse circulation in the immediate area of the junk basket is thereby created.

Caution should be observed in dropping or circulating any ball or other restriction plug down the drill pipe or tubing. As noted previously, it is most necessary to caliper, measure, and note all dimensions of all tools that are run downhole. Some have restricted inside diameters and will not allow balls or other items to pass. It should be standard practice to check this carefully before dropping anything down the work string.

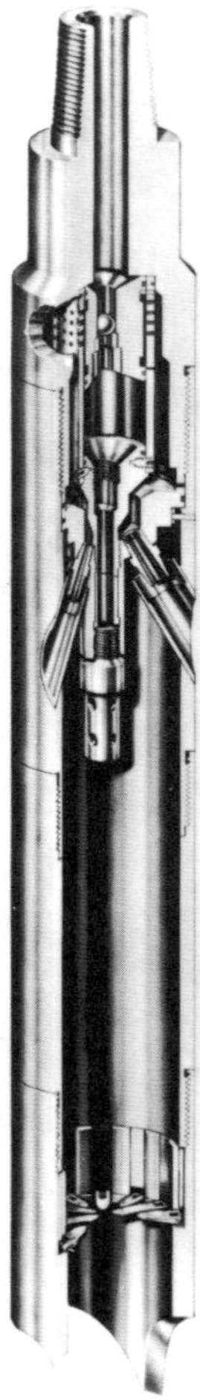

Figure 10-4. Jet-type junk basket. (Courtesy of Houston Engineers, Inc.)

Friction Sockets

Many times the standard manufactured junk basket will not adapt to the particular problem due to the size and shape of the junk in the hole. Ingenuity needs to be used to devise alternatives and adapt to the problem at hand. If the I.D. of the catcher is not large enough to accommodate the junk, a shoe or length of pipe may be used as the body of a shop-made junk basket. A series of holes may be bored or burned around the circumference of the material and then steel cables brazed into place forming a catcher. It is not possible to rotate this tool on the junk as the cables will be broken and torn out but the tool can be pushed down over junk and the

Loose Junk Fishing 61

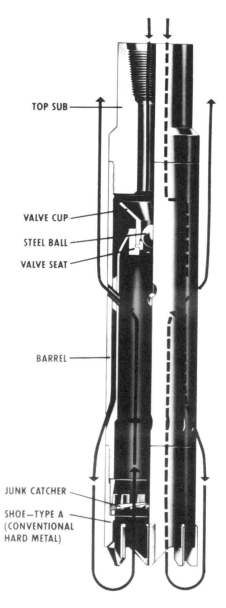

Figure 10-5. Reverse circulation junk basket. (Courtesy of Bowen Tools.)

junk retained by the catcher. A friction catch can also be made by cutting inverted "U's" in a piece of pipe and bending them in until they practically touch. This tool can be pushed down over a long tubular piece of junk and is quite effective in cases where the dimensions are unknown.

62 Oilwell Fishing Operations

One of the foregoing two designs can remedy the two problems that are most prominent—the junk is too large to catch and the outside diameter is not known.

Several good "mouse trap" design tools have been made in past years, but unfortunately they were used mostly by cable tool operators and have not remained available to the rotary tool market. One design has tracks set on opposite sides and at an angle from bottom to the top of the bowl. (See Figure 10-6.) Various slips can be fitted to ride up and down on the beveled track. As the tool is lowered over the fish, the slip is pushed up until sufficient clearance has been obtained for the fish to pass the slip. The slip falls down and wedges the fish in the bowl. This tool cannot be released, but it is very effective for fishing sucker rods in casing or tubing that is so corroded that an ordinary overshot will not catch it.

Poor Boy Basket

These were probably the very first designs of junk baskets. They were used by early drillers and cable tool operators before such tools as are available today were manufactured.

The poor boy basket is usually shop made for the particular job from a short section of low carbon steel pipe. Schedule 40 material is a good choice but anything of higher grade than J-55 will not work properly, as the teeth will break off without bending.

Note in Figure 10-7 that the basket is made with teeth cut with a welding torch and with a curved leading edge. This edge is also cut with a bevel. Note also that there is a gap between the teeth and that the teeth are about three-fourths the diameter of the pipe from which they are made. This length of pipe is then threaded or welded to a top sub or bushing.

The running of the tool is most important. It is necessary to rotate and circulate the tool down over the junk without excessive weight. Due to the slots between the teeth, the tool will usually run rough while the junk is at the level of the teeth. As hole is made (by measurement) and the junk moves up into the smooth bowl of the basket, the tool will run smoothly. When this has been accomplished, weight is applied as the tool is rotated and the fingers will bend in and "orange peel," retaining the junk inside the bowl. New teeth must be cut for each job.

Boot Basket

This tool is also called a "junk sub or boot sub" (Figure 10-8). It is made to be run in conjunction with and just above some other tool such as a bit, mill, magnet, or catcher-type junk basket. It will operate properly

Loose Junk Fishing 63

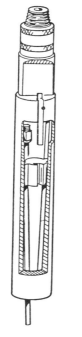

Figure 10-6. Clulow socket: slip has wedged a sucker rod in bowl.

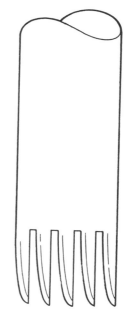

Figure 10-7. Poor boy junk basket.

only when the circulation is the "long way" or down the work string and up the annulus.

The boot on the basket is comparatively large for the hole or casing size (Table 10-1), therefore high-velocity fluid return is accomplished through this portion of the string. As the fluid reaches the top of the boot, a much greater annulus area is present and the fluid pressure is dropped suddenly, creating a turbulent flow just above the top of the boot. Any heavy particles such as steel cuttings, carbide inserts, bit teeth, or ball bearings that are circulating in this fluid will tend to drop at this point and fall into the boot. Boot baskets may be run in tandem to increase the capacity, and some operators will also place another boot basket up the hole several joints to pick up junk that has been pumped higher than the lower basket.

Field welding should not be permitted on the mandrel of the boot basket. Operators have welded gussets on the mandrels to reinforce the boot; but without stress relief, these welds may produce stress cracks resulting in the failure of the mandrel and in turn an expensive fishing job.

Table 10-1
Boot Basket Recommended Sizes

Hole Size or Pipe I.D. (in.)	Boot O.D. (in.)	Connection API Reg.
$4^{1}/_{4}$–$4^{5}/_{8}$	$3^{11}/_{16}$	$2^{3}/_{8}$
$4^{5}/_{8}$–$4^{7}/_{8}$	4	$2^{7}/_{8}$
$5^{1}/_{8}$–$5^{7}/_{8}$	$4^{1}/_{2}$	$3^{1}/_{2}$
6–$6^{3}/_{8}$	5	$3^{1}/_{2}$
$6^{1}/_{2}$–$7^{1}/_{2}$	$5^{1}/_{2}$	$3^{1}/_{2}$
$7^{1}/_{2}$–$8^{1}/_{2}$	$6^{5}/_{8}$	$4^{1}/_{2}$
$8^{5}/_{8}$–$9^{5}/_{8}$	7	$4^{1}/_{2}$
$9^{5}/_{8}$–$11^{3}/_{8}$	$8^{5}/_{8}$	$6^{5}/_{8}$
$11^{1}/_{2}$–13	$9^{5}/_{8}$	$6^{5}/_{8}$
$14^{3}/_{4}$–$17^{1}/_{2}$	$12^{7}/_{8}$	$7^{5}/_{8}$

Hydrostatic Bailer

When cleaning out miscellaneous junk in a wellbore, it may be practical to run a hydrostatic bailer (Figure 10-9). Different designs are made for running on pipe as well as sand lines. All bailers work on the hydrostatic head principle, as they depend on the weight of the fluid in the hole to force the junk into the bailer and past the catchers. Many bailers can be surged repeatedly until the basket is full of junk or the hole is clean. They are particularly appropriate for cleaning out bit cone parts, bearings, pipe slivers, bolts, nuts, perforator debris, and other material that is nonmagnetic.

Rotating Bailers

A new concept in downhole remedial and completion tools has given rise to bailers which can be rotated for drilling or milling (see Figure 10-10). There are several designs of these tools on the market but all operate under the same basic principle. This type of tool consists of component parts which are assembled in the work string at the well site. The complete bailer is made up of the appropriate tool on the bottom, such as a bit, mill, retrieving head, or rotary shoe, and several check valves, a reciprocating pump and kelly, and if applicable, a perforated sub.

The heart of the tool is a plunger pump similar to that used for pumping oil wells, with tailpipe (a cavity and a series of check valves) run below the pump. By reciprocating the tubing at the surface, the pump pulls fluid and debris (sand, scale, cement, junk, etc.) up into the tailpipe where it is held by the check valves. The fluid may be pumped to the surface or recirculated into the annulus through a perforated sub positioned above

Loose Junk Fishing 65

Figure 10-8. Boot basket. (Courtesy of Gotco International, Inc.)

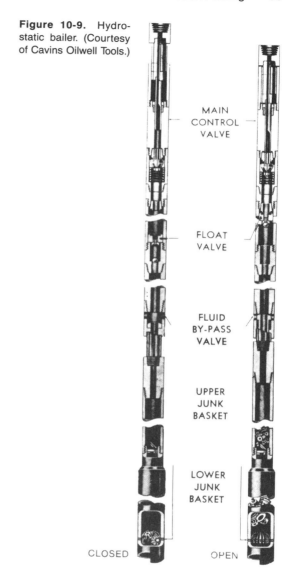

Figure 10-9. Hydrostatic bailer. (Courtesy of Cavins Oilwell Tools.)

the pump. This feature allows the tool to be operated in low fluid level wells with the same efficiency as in a full hole. When the tailpipe is full of settled out debris, the work string is tripped and the fill is emptied.

The plunger pump incorporates a kelly and drive bushing in its design. The tool can be used with a power swivel for drilling and milling of such items as cement, scale, sand, packers, plugs, and junk.

66 Oilwell Fishing Operations

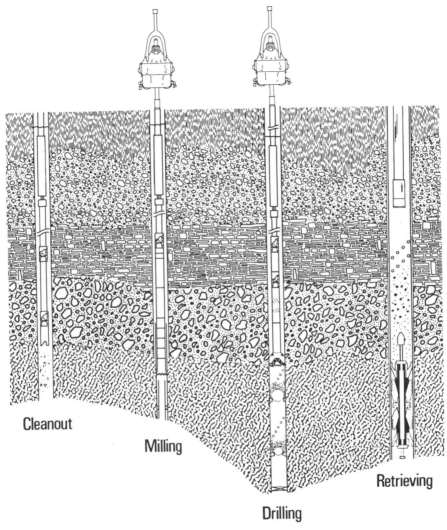

Figure 10-10. Rotating Bailer Applications. (Courtesy of the Adaptable Tool Co.)

No Formation Damage

Fluid loss to the formation is kept at a minimum because the tools work under the normal hydrostatic pressure in the well. This helps prevent formation damage and water blockage. The savings can be tremendous. The costs of transporting the fluid, filling the well, producing the fluid later, treating it, and disposing of it are all eliminated.

Junk Shots

Junk shots are comparably large jet-shaped explosive charges (Figure 10-11) (run on either drill stem or electric wireline) to break up objects left in the hole which are not recoverable by ordinary fishing methods. The charge breaks up the junk into small pieces which can be recovered usually with magnets or junk baskets. Since the large size of the charge creates a tremendous force, a cavern may be created and sometimes parts of the debris are blown out into the sidewall of the hole. All of the force of the explosion cannot be directed downward even though the tools are designed so that the maximum force is in this direction. The target distance with any shaped charge is very critical; therefore for maximum efficiency the charge should be circulated down to the fish if the junk shot is run on drill pipe. If the shot is run on an electric wireline, then a bit run should be made to ensure that the charge gets completely down to the fish.

A junk shot should never be run inside pipe, as the explosive force will destroy the casing or pipe and probably increase the problems.

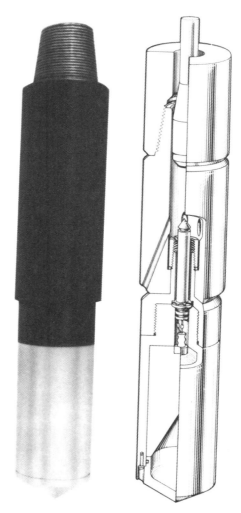

Figure 10-11. Junk Shot. (Courtesy of Well Control.)

Chapter 11
Tungsten Carbide Mills and Rotary Shoes

Cutting and milling tools dressed with tungsten carbide are probably the greatest innovation to be adopted in the fishing tool business in the past thirty years. Prior to the current designs and manufacture of tungsten carbide tools, "fluted" mills and shoes were used. These tools had cutting teeth or blades that had been carburized, i.e., the outer surfaces had been hardened for cutting while the inside metal was still in a semi-annealed state and somewhat resilient by comparison. This prevented breaking of the teeth or blades.

Material

Tungsten carbide material for the dressing of mills and shoes is sold in sticks or rods approximately 18 in. long. The particles of sintered tungsten carbide are very irregular in shape and have sharp edges like broken glass. These pieces of carbide are imbedded in a matrix material of nickel-silver bronze. In any single stick all of the carbide particles have been screened according to size and they may be graded, for example, as 3/8 in.–1/4 in. or perhaps 30–45 screen mesh. For effective cutting the carbides must be of good quality and perfectly clean, since any dirt, oil, or trash will prevent their sticking to the alloyed bronze matrix material. Normally large particles are used for larger O.D. mills and shoes, while the smaller particles are used on small tools, wear surfacing of tools such as reamers, and the filling in between the larger particles in large tools. More cutting surfaces produce more steel cuttings.

There is a significant difference in the quality of carbide sticks that are available. The complete coating of the carbide particles with the alloyed bronze in the finished mill or shoe is very necessary. It is much easier for the welder if the stick already contains particles that are completely coated. The matrix material is resilient and helps to withstand shock and sudden loads. It has an ultimate shear strength of approximately 100,000 lb/sq in.

Manufacture or "Dressing"

The application of the tungsten carbide to the tools is a brazing process using oxy-acetylene equipment and practices. Since the gases generated by the melting of the bronze are toxic, the welder must work in a well-ventilated location with sufficient draft to remove the fumes.

The mill body or blank rotary shoe should be thoroughly cleaned or machined just prior to the application. Any grease, dirt, or rust will prevent a strong bond. The body of the tool is preheated, and the surface to be dressed is tinned with a tinning rod for good bond. The welder flows the material on to the body by melting it with the torch and puddling it with the stick of carbide. This process is a very painstaking operation and requires considerable time (Figure 11-1). A six- or eight-inch diameter mill can easily require six or eight hours of work to dress properly.

As the welder applies the carbide to the body of the shoe or mill, he shapes the tool into the configuration that is desired. As shown in the illustrations, there are many designs of carbide tools for particular applications and the welder applies the carbide to accomplish the proper design. The welder may use "heat sticks" to estimate temperatures attained so that there is no damage to the carbides by excessive heat, yet a temperature is reached that assures good bond of the matrix material. The finished tool is never quenched but is allowed to cool slowly. In larger sizes it is sometimes wrapped in insulation.

Design

Carbide tools should be carefully designed for each particular job. If a mill is run on a work string (i.e., not washpipe), there should not be a circulation hole in the center. Such a design will tend to cut a core and then have a tendency to spin on the core. One of the blades should be long enough to extend past the center. In the design of a carbide rotary shoe, internal build-up of carbide is beneficial if there is sufficient clearance. This build-up will trim the fish down to a size that will allow it to be swallowed in the washpipe and not plug it off, bind it, or twist it off. Many times when there is a substantial shoulder inside the rotary shoe, the fish or portions of it will be brought out in the shoe and washpipe without the use of any retrieving tool.

70 Oilwell Fishing Operations

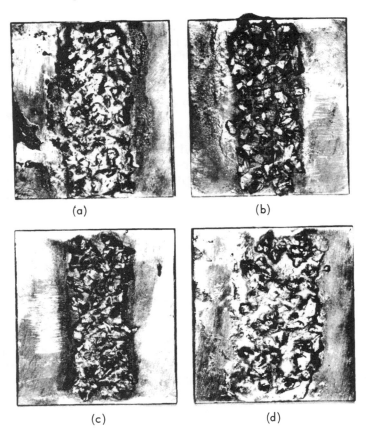

(a) (b)

(c) (d)

Figure 11-1. Carbide application (unretouched photographs of samples): (A) Correct application. This sample shows the correct application at proper heat. The matrix is well bonded to the base metal. The tungsten carbide particles are compactly spaced and securely imbedded in the matrix material. The resulting application, when cool, has a slightly golden hue. (B) Improper application. This sample shows the result of too much heat. The heat has dissolved the matrix material. The tungsten carbide particles are burned and charred. The resulting application, when cool, has a black and burned appearance. (C) Improper application. This sample shows the result of too little heat. The matrix is not bonded to the base metal. Although the tungsten carbide particles are imbedded in the matrix, the material will chip and break away from the base metal when milling. The resulting application has a dull silver appearance. (D) Improper application. This sample shows the result of improper manipulation and spacing of the tungsten carbide particles. Although applied with proper heat and well bonded, large vacancies exist and the result would be an inefficient milling surface. In appearance, the resulting application would have the slightly golden hue as in Figure 10-1A, but the vacancies or cavities would be very apparent. (Courtesy of Bowen Tools.)

Tungsten Carbide Mills and Rotary Shoes   71

Carbide should never be left on the outside of a shoe or mill that is to be run in casing. It should be ground smooth and made concentric with the body of the tool to prevent any damage to the casing or pipe. In some deviated wells, bronze wear pads are made on the outer portion of the shoe to stabilize it, to reduce friction and torque, and to hold the cutting portion of the shoe away from the casing. Tools that are to be run in open holes are dressed with carbide on the outside, so that they continue to cut on any junk that gets up alongside the shoe or mill. (See Figures 11-1 through 11-8.)

Running Carbide Tools

Running a carbide tool properly is just as important as making a quality tool. Many tools are damaged before they have a chance at cutting prop-

Figure 11-2. Junk mills. These may be designed with blades for drilling cement and steel or cement and such equipment as float shoes. Note that the hole is not centered. Mills for use in casing are ground smooth on the outside circumference. For use in an open hole, the mill would be dressed with carbide on the outside of the head. (Courtesy of Petco Fishing & Rental Tools.)

Figure 11-3. Flat bottom mills. This particular design is quite limited in use. If it is to be used in an open hole, it should be dressed with carbide on the sides of the head. It may be advisable to have a concave face on the bottom as this would tend to keep the junk centered. Note the pads on the sides, which stabilize the tool in the hole. (Courtesy of B & W Metals Co., Inc.)

erly and making hole. Carbide milling tools are very similar to the cutting tools that a machinist uses in cutting steel in a lathe or other machine tool. They should be run with the same general rules and procedures. Oil field operators are accustomed to running rock bits which require considerable weight to be effective. As one can see, the weight on a rock bit breaks up the rock with the teeth as the cones rotate. The cuttings are then washed away by the drilling mud or workover fluid. Since the mill or shoe is basically a cutting tool, the procedure should follow that used on a machine tool. Rotation is fast in order to secure sufficient linear speed. A 4-in. diameter mill should be rotated at approximately 175 rpm and a 12-in. diameter mill or shoe should be turned at approximately 60 rpm. The fast speed creates a cutting action and prevents grinding, which is destructive to the mill or shoe. The pump should be turned on and circulation established prior to the carbide tool touching the fish. It is important that cuttings are washed away immediately, as they tend to "nest" or ball up and plug the tool or annulus. To lift cuttings, fluid viscosity should be 50 to 80 cp depending on the size of the cuttings. The tool should be lowered very slowly and a determination made in the first 30 minutes of how fast one can slack off on the work string to effectively make hole. Excess weight will merely grind away at the carbides and matrix. The tool will develop excessive heat with too much weight, and results will be very disappointing. Always keep in mind that the cutting action is similar to that of a fly cutter in a machine tool.

If the cuttings cannot be brought to the surface with the circulating fluid, boot baskets must be run just above the mill or drill collars to catch the cuttings so that they don't bridge or plug the hole. It is sometimes possible to reverse circulate. This will bring cuttings to the surface due to the higher velocity in the smaller cross-sectional area of the work string. However, plugging of the mill and pipe with the steel cuttings can also

Figure 11-4. Tapered mills. This design can be used to open the top of a liner or to enlarge a hole that has been started. It is not a desirable tool to use in collapsed pipe, as there is a tendency to follow the collapse and go outside the pipe. (Courtesy of Petco Fishing & Rental Tools.)

74   Oilwell Fishing Operations

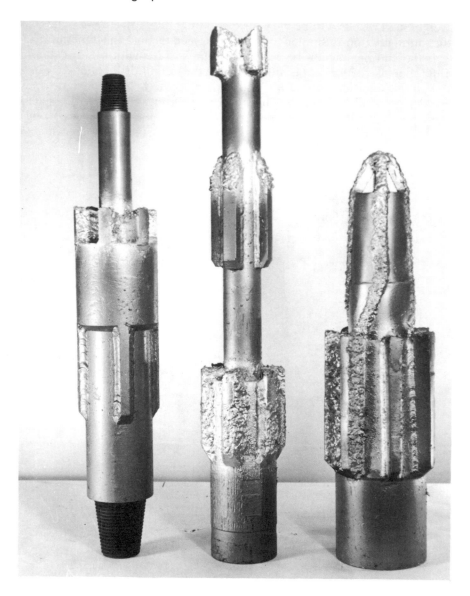

Figure 11-5. Pilot mills. Typical designs to mill up pipe in the hole. Design on the left is appropriate for milling up a permanent packer of small diameter. Stinger would be extended with a retreiving tool on bottom. (Courtesy of Petco Fishing & Rental Tools.)

Tungsten Carbide Mills and Rotary Shoes 75

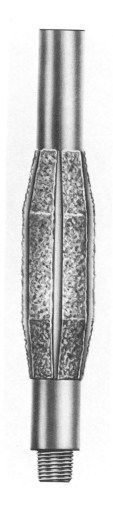

Figure 11-6. Watermelon or string mill. These are used to open up tight spots in pipe, to enlarge and clean up a window cut in casing, or, under certain circumstances, to run in collapsed casing that has been partially opened up. With a guide below the mill, it will follow the pipe or hole and will not go outside as will a tapered mill. (Courtesy of B & W Metals Co., Inc.)

Figure 11-7. Blade reamer. These are used in holes to keep them in gauge. Carbide may be used on the blades as cutting material for hard formation and to build up reamers that have worn down out of gauge. (Courtesy of B & W Metals Co., Inc.)

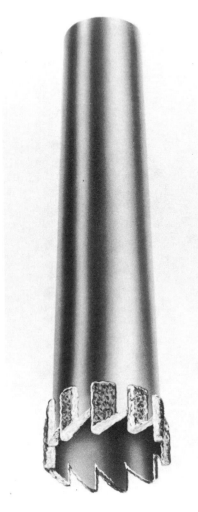

become a problem when reversing. Ditch magnets are frequently run in the return line at the surface to collect any steel cuttings that have passed the shale shaker screen. This helps to estimate the amount of material that has been milled up, and it also prevents cuttings from getting in the pumps and damaging them.

Carbide mills and shoes may be redressed if they are properly prepared. The old carbide must be washed off any debris that is imbedded in the matrix, and the material must be perfectly clean to ensure a good bond. The same precautions followed in the original dressing must be taken in redressing a worn tool.

Figure 11-8. Tooth-type rotary shoe. Carbide is frequently used on rotary shoes for any application where metal cutting may be necessary. There are many designs fo rotary or "burning" shoes. The same rules should be followed in selecting rotary shoes as when selecting carbide mills. Do not use carbide on the outside when running inside pipe. Dress the inside with carbide if there is enough clearance so that the fish will be trimmed down to prevent plugging and excessive torque. (Courtesy of B & W Metals Co., Inc.)

Chapter 12
Wireline Fishing

One of the most challenging of fishing jobs may be the recovery of wireline and the tool or instrument run with it. First, situations must be separated between those where the line is still intact and situations where the line is parted. We shall also differentiate between electric or conductor lines and swab and sand lines.

If a conductor line has not parted, good practice usually dictates that we should not pull out of the rope socket. This causes us to lose contact with the tool or instrument, and it may become permanently lost. If the instrument contains a radioactive source, the situation becomes even more critical.

The operator has the choice of using either the cable-guide method (better known as "cut and strip") or the side-door overshot method. The cable-guide method should be chosen for all deep open-hole situations or when a radioactive instrument is stuck in the hole. This is the safest method and assures a very high success ratio. The only disadvantages are that the cable must be cut and the stripping-over procedure is slow and time consuming.

Cable-Guide Method

A special set of tools is required and these are usually kept by the fishing tool service company in a special box or container, since they are not used for other purposes.

The tools (Figure 12-1) consist of a cable clamp with "T" bar, rope sockets for each end of the line, a sinker bar, and special quick connector-type overshot for the line on the reel end and a spear point for the well end. There are also included a slotted plate to set on top of the pipe, a sub with a recess or retainer to hold the rope socket, and an overshot to run on the pipe to catch the instrument or tool.

78 Oilwell Fishing Operations

A slight strain (approximately 2,000 lb) is taken on the line and the cable hangar or clamp (Figure 12-1A) is attached to the cable at the well head or rotary table and the cable lowered so that the hanger rests at the surface. The cable can then be cut at a convenient length above the floor. Caution should be used to allow enough length. As in any deviated hole, the cable is pulled out from the wall, and more length is required to reach the surface than before it was stripped inside the pipe. Rope sockets are then made up on each end of the line with the overshot (E) on the upper end and the spear head (B) on the lower end. As each stand of pipe is run, the cable spear head rests on the "C" plate (F) which prevents the line from falling.

The first stand of pipe (Figure 12-2) to be run is made up with an appropriate overshot on bottom to catch the rope socket, fishing neck, or body of the tool in the hole. Caution must be exercised to ensure that the guide or bottom does not have any sharp edges that would cut the line if pipe weight were set down on it in a dog-leg or on a ledge.

The line to the reel is spooled up to the derrick man who stabs the spear head overshot and sinker bar in the pipe. With the pipe hanging in the

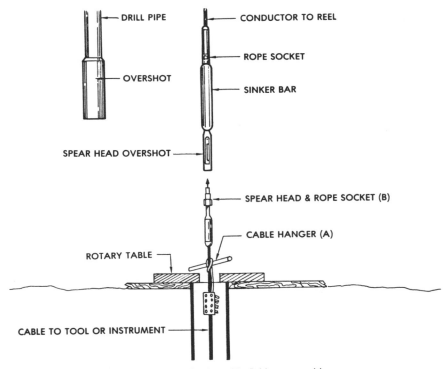

Figure 12-1. Cable guide fishing assembly.

Wireline Fishing 79

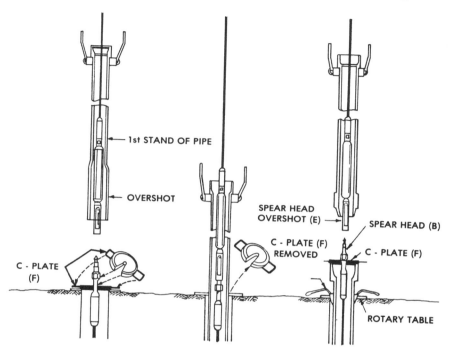

Figure 12-2. Cable guide fishing method.

derrick, the spear head overshot is lowered through the pipe to the floor man who connects the spear head overshot with the spear point. The line is then picked up, and the stand of pipe can be run. This procedure is repeated until the overshot has tagged the body or fishing neck on the tool so it can be engaged. Concern at this time is the proof that the instrument is truly caught. The first check is to pick up the pipe and the line should be slack. There is a sub fitted on the top of the pipe which has a restriction or "side pocket." The spear point rope socket may be set into this recess and the kelly or fittings made up on the pipe. This will allow pump pressure to be exerted against the fish in the overshot to ensure that it is safely caught and will not fall out coming out of the hole.

After tests have been made to show that the fish is securely caught, the clamp may again be placed on the line below the cut portion, the rope sockets removed, a square knot tied in the two pieces of line, and the line pulled out of the rope socket with the elevator and the clamp. The line may be spooled up and the pipe with the instrument or tool recovered.

As with all tools run in a well, wireline tools including rope sockets, fishing necks, and instrument bodies should all be measured or calipered before running. If fishing these tools by the preceding method, the overshot above the grapple must be sufficiently open to swallow anything above the part that is being caught.

Side-Door Overshot

The side-door overshot (see Figure 12-3) is a special overshot having a gate or door in the side which can be removed to allow the line to be fed into the tool, after which the door is put back into position as part of the bowl. The overshot is run on the drill pipe or tubing until the fishing neck or body of the stuck tool is engaged and caught.

The advantage of this method of recovery is that the line need not be cut. It is also fast since no stripping is necessary. Care must be exercised in setting the slips with the cable or line in the gap to prevent pinching or cutting. Since the line is outside the pipe, care must be taken not to rotate the pipe excessively, as this wraps the line around the pipe.

Side-door overshots are not run in deep open holes because the line frequently becomes key seated and even differentially stuck in the filter cake. All open holes are crooked enough to cause the line to drag the side wall and cut a groove in the filter cake, causing excessive friction and sticking.

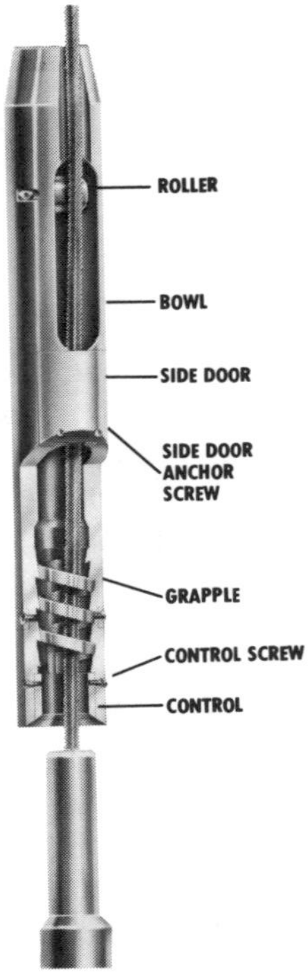

Radioactive Sources

In recent years there has been a tremendous increase in the use of radioactive sources in well logging, with the attendant increase in agencies that regulate their use. Many operators do not have a clear understanding of the regulations in effect when a source becomes lodged or lost in the hole. The source should be recovered intact if at all possible from an economic viewpoint or it may be abandoned if such is possible to protect personnel and property in the future.

If it appears that the source cannot be recovered, the logging company must notify the Nuclear Regulatory Commission and the state regulatory body. Plans are then discussed and a decision made which can be approved by the agencies concerned. If the source is abandoned in a dry hole, all records, including those of the agency that issued a permit for drilling the well, should contain complete information and the well head, or a suitable marker, should contain full information on an engraved metal plate.

Figure 12-3. Side door overshot. (Courtesy of Bowen Tools.)

When a source is left below a producing zone, usually the normal cementing of the casing will isolate the source sufficiently. Most capsules that contain radioactive sources are made to resist corrosion and erosion for many years. Abrasion from any fluid flow with its attendant sand or other particles is the greatest hazard. In order to gain approval for the abandonment of the source, a plan must give assurance that there is no reasonable probability of fluid flow past the capsule or that it will not be encountered if the hole is sidetracked.

Fishing Parted Wireline

Wireline does not fall down the hole when it parts. Of course it falls, but not like a hemp rope or chain. It is surprising how high it stands. The larger the line, the stiffer it is. The smaller the diameter of the pipe or hole, the less the line can fall. Since both of these factors vary a great deal, it is not possible to use a rule of thumb except to suspect that the top of the line is higher than anticipated.

Rope Spears and Grabs

The rope spear or center spear (Figure 12-4) is the most desirable tool for fishing wireline. Each spear must be adapted to the size of the hole or pipe and the barbs checked to be certain that the line will wedge in them enough to pull the line in two if necessary. If the tool is run in pipe, then a stop should be run just above it. The stop must be of sufficient size so that the line will not go above it. This prevents catching the line low, balling up a long length, and sticking the fishing string.

Always try to catch the line near the top. The more line that is pushed down the hole, the more compacted it becomes and the more

Figure 12-4. Center spear or rope spear.

difficult it is to penetrate and catch with the barbs on the spear.

When it is no longer possible to make a good catch with the center spear, a two-prong grab (Figure 12-5) is usually run. This permits catching the wireline from the outside instead of the inside. Again, one should be certain that the barbs will wedge the line and that there is sufficient clearance inside the tool between the prongs or barbs for the line. After a new top is secured with a run with a two- or three-prong grab, the center spear is again run, since it is considered to be the safest.

If the rope spear is run below casing into the open hole, the stop is not run, because the open hole would not always be in gauge and the line could pass the stop, stack up on top of the stop, and prevent pulling back into the pipe.

Box Taps

If the line becomes packed and it is impossible to penetrate it with either the center spear or two-prong grab, it may be possible to screw a box tap (Figure 12-6) on the ball of the wireline. This should be full O.D. with a thin wall near the bottom. After the ball of line is fully engaged by the box tap, the line can be pulled in two, which provides a new top.

Always keep up with the amount of line that is recovered. Since it is usually unbraided, balled up, and no longer the original length, weight may be the only method of estimating the amount recovered and, in turn, the amount remaining in the wellbore. If this is true, it is necessary to determine the weight of the line per foot and weigh what is recovered.

Figure 12-5. Two-pronged grab.

Since the line stands up in the hole, if the length left on the tool is short, even 100 ft, it may be possible to fit an overshot to catch the tool and swallow the line. The overshot and the extensions and pipe above it

Figure 12-6. Box tap. (Courtesy of Gotco International, Inc.)

should not have any small restrictions or square shoulders. The overshot is slowly rotated as it is lowered. Short lengths of line can be caught complete with the tool, which is much easier than fishing for the line itself.

Cutting The Line

When a sand line or swab line is stuck in the hole, it is usually advisable to cut the line as low as possible so that the wireline can be recovered and the tool fished with a work string of pipe. This is also advisable on some occasions when a conductor line is run inside the pipe and the tool becomes stuck.

Figure 12-7. Sand line cutter (rigged with sleeve for 2⅞-in. O.D. pipe and larger). (Courtesy of J. C. Kinley Co.)

84 Oilwell Fishing Operations

In the early days of cable tools, a rope knife was stripped in over the stuck line and run on another line. Due to the lay of the two lines, the second line frequently became stuck creating an even more serious situation. This method has now become almost extinct with the advent of the explosive sand line cutter (see Figure 12-7).

Currently the most popular cutter is a cylindrical tool long and small enough in diameter to be run inside 2-in. tubing and to cut a 9/16-in. line. Even smaller tools are available through special order. The cutter is dropped around the line. It is grooved so that it rides the line down to the top of the rope socket. It is then fired by sliding the drop weight down the line onto it. A small propellant charge drives the wedge which forces the knife to cut the line (Figure 12-8). In the smaller tubing, the drop weight and the gun are both provided with a fishing neck which can be used to recover them by means of a pulling tool on a measuring line. In 2⅞-in. tubing, or in drill pipe, casing, or an open hole, a sleeve and sometimes guides are installed on the tool. The sleeve provides a seat for the cutting edge, and it also allows a crimper to be installed so that the drive wedge, which operates the cutting knife, also forces the crimper to clamp the line against the adapter sleeve. Then the gun and drop weight can be recovered together on the end of the cut line. Since this cutter is a free-falling tool, it is advisable to work the line to ensure that the gun falls as deeply as possible. This can be done by taking a strain on the line, releasing it, and letting it fall five or six feet before catching it with the brake. Shaking the line in this manner will work the gun past some obstructions. However, the gun will stop on some splices, flags tied on the outside of the line, or in mashed tubing. Wherever it stops, the cut will be made. There is no way of knowing where the gun cuts until the cut line is spooled up.

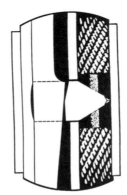

Figure 12-8. Cross section of knife of sand line cutter. Wedge has driven knife through line. (Courtesy of J. C. Kinley Co.)

The sand line cutter is also available with an electronic timer programmed to fire the tool after a timed interval. It is used in wells where falling sand or solids tend to cover the gun and prevent the drop weight from hitting the firing pin. It is also used in deviated wells where the drop weight speed may be too slow to fire the cutter.

Electric Submergible Pumps

When sucker rod pumps and downhole hydraulic pumps can no longer lift sufficient fluid, electric submergible pumps are used. They consist of an electric motor, a pump, and usually some device for gas separation. In order to power the electric motor, it is necessary to run a three-phase electric conduit down to the motor. This cable is usually strapped to the production tubing with stainless steel packing crate-type straps.

In order to contain a powerful electric motor and a pump, the housings of electric submergible pumps are relatively large in diameter and do not have much clearance between the pump and casing.

Electric Submergible Pump Data

Casing O.D. (in.)	Pump O.D. (max.) (in.)	Bhp at Pump Shaft (max.)	Fluid Bpd (max.)
4½	3⅜	125	1,900
5½	4	250	4,500
7	5⅛	480	9,000

Small deposits of sand, corrosion, gyp, etc. between the casing and pump will stick the pump so that it can't be retrieved. Care should be exercised so that the tubing is not pulled in two as the cable, unlike other wireline, is flexible and quite heavy. If it parts, it falls and is easily packed down so that a center spear or other tool cannot penetrate it to retrieve it without its parting into short lengths. Figure 12-9 illustrates how the cable may become packed in the casing if the tubing is parted and pulled.

When pumps or the tubing become stuck, the tubing string should be free-pointed and the tubing chemically cut above the stuck point. Most installations include a check valve in the tubing string to prevent a back flow of fluid when the motor is turned off, causing damage to the motor. Free-point and cutting tools cannot be run below this check valve, so it is recommended that the valves be placed as low as possible.

It is most important that only a chemical cut be made to part the tubing string, as this leaves a sharp cutting edge on the tubing, which is used as a knife to cut the electric conduit.

86   Oilwell Fishing Operations

Figure 12-9. Electric submergible pump cable packed in casing.

When the chemical cut has been made, the tubing will be parted but the conduit (cable) will still be intact. The tubing is then raised 18–24 in., providing a gap at the tubing cut with the cable (conduit) pulled taut between the two sections of tubing.

The electric wireline tool (Figure 12-10) is made up with a sinker bar and bumper jar and run on a work string of sucker rods or small tubing.

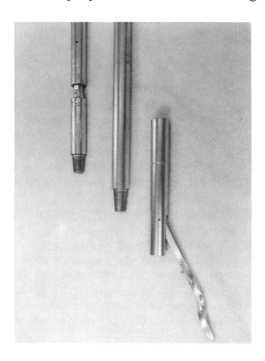

Figure 12-10. Electric line tool (Patent No. 4306622), shown with sinker bar and bumper jar. (Courtesy of Armstrong's Oilfield Service Co., Inc.)

The cutter is equipped with a spring-loaded arm which extends from the mandrel. When the cutter is measured in to the depth of the gap, the arm extends, and as the tool is rotated to the right, the arm catches the cable and pulls it up against the mandrel. By striking a series of blows up with the bumper jar, the cable is severed by the sharp edge of the chemical cut. It can be easily determined when the cable is cut as torque in the work string will be lost. When the cable is cut, the work string with the electric wireline tool and then the production tubing string with the electrical cable strapped to it can be pulled.

This will leave the pump and short section of tubing and cable in the well. These should be fished with an overshot or spear and a jarring string.

Depending on the design of the electric submergible pump, caution should be used in jarring upward as some designs incorporate a flange on top which can be easily parted. Light blows should be used, both up and down, until some travel in the fish is accomplished. When this occurs, continued movement of the pump will work it free.

Chapter 13
Retrieving Stuck Packers

While it is true with any fish, it is especially helpful in fishing stuck packers that we know as much about the equipment as possible. If we know the make and designation of the packer, we can obtain complete data such as major dimensions, type, and method of setting and releasing, plus a picture or dimensional drawing. Many times, it is most helpful to take another packer of the same type to the well location. This provides immediate information if only a portion of the equipment is recovered and other parts are left in the well. Old catalogs should never be discarded, as manufacturers change designs, sell out, change names, and go out of business.

Packers generally fall into two main categories: retrievable and permanent. Packers in the retrievable design include weight-set (with either "J-Set" or "automatic bottom"), tension type, and rotation set (with alternate weight and pick-up on the setting string). Some packers have hydraulic hold-downs above the seals so that no tubing weight has to be left on the pipe. Others are set hydraulically and must be released by rotation or shear pins or rings.

Permanent packers are simple in comparison to retrievable packers. They consist of slips at the top to prevent the packer moving up the wellbore after setting, a seal, and a set of slips at the bottom that prevents the packer from moving down the hole. This configuration may be modified by adding seal bore extensions and equipment such as blast joints below the packer.

Retrievable Packers

After it is determined what type packer is in the well, efforts should be methodically carried out to release it or if necessary, to fish it. The tubing should be worked to ensure that it is completely free and that pipe friction is not adding to the problem. If the packer is released by rotation, the torque must be worked down. The pipe is marked with a vertical mark, and right-hand torque is applied at the surface. While holding this torque, the pipe should be reciprocated. This ensures that the torque is distributed and that some torque is applied to the mandrel. Stretch should be measured and an estimate made of the depth of the highest stuck point. Running of a free-point instrument would also be considered at this point. If it is found that the slips are frozen, it is sometimes profitable to fire a string shot in the packer mandrel. If the formation will permit, pressure can be applied down the tubing and below the packer to provide lifting force on the packer. A hole may be punched in the tubing just above the packer mandrel and the wellbore circulated in case there are solids that have settled out in the annulus. If the packer is equipped with a hydraulic hold-down above the seal and slips, pressure may be applied to the annulus to help retract the hold-down buttons.

A jarring string is usually very effective if the retrievable packer itself is stuck. The tubing string is parted by either cutting or back-off, and the appropriate catch tool is run with the jars, as described in Chapter 8.

The alternative method of retrieving a packer would be to wash over it and cut it out if necessary. This would probably be the chosen method if some of the tubing is stuck due to fill in the annulus. If only a short section of tubing is stuck and it is practical to wash over it in one trip, a dog overshot may be incorporated in the washover string. These overshots consist of a short section of washpipe (bushing) made up in the string and having an appropriate internal catcher to engage under the couplings of the tubing as in an external cutter (Figures 9-5 and 9-6). Rotary shoes for this operation would be of the tooth type for digging out fill, mud, or cement and would incorporate carbide if the packer itself should have to be cut.

Permanent Packers

Cutting over and retrieving permanent packers is a very common job, and it can be a very efficient procedure. The tool string consists of a carbide rotary shoe or mill, top sub or bushing (when appropriate), a length of small pipe for use as a stinger, and a releasing spear or retriever. The tools are made up as shown in Figure 13-1. Take care that the stinger is made up tightly and then pinned, set-screwed, or strapped so that it will not back off. Note that the right-hand rotation of the assembly tends to back off the stinger. Friction of its running in the bore of the packer provides back-up, hence the tendency for this part of the tool assembly to back off.

If the packer is small such as in 4½-in. O.D. casing, a mill should be selected as a shoe of this size would be weak; it would tend to wedge and

90 Oilwell Fishing Operations

Figure 13-1. Permanent packer milling and retrieving tool. (Courtesy of Houston Engineers, Inc.)

very little savings would be made in the amount of material to be cut. If the packer is in larger casing, an appropriate shoe should be selected, dressed on the bottom, with a small shoulder of carbide on the inside to provide clearance as the shoe cuts over the packer. The outside of the shoe or mill should be perfectly smooth so that it does not cut the casing. The shoe should be long enough to cover the entire packer and if necessary an extension provided.

Retrievers are made in several designs, but most can be operated through a "J" on the mandrel with springs to provide back-up for operation. The retriever is run in the retracted position and is small enough to go through the bore. It is then set so that the slip, grapple, or catching device is extended and it will not come back through the bore. With the catcher extended, a pull on the work string exerts a pull on the bottom of the packer.

When sufficient "hole" has been made to ensure cutting through the top slips, rotation and pumping should be stopped and a pull exerted on the packer. Frequently, this is enough cutting to free the packer for travel upward and out of the wellbore.

If equipment such as blast joints, sliding sleeves, etc. are run below the packer, a section of pipe of an I.D. larger than the packer bore is usually run immediately below the packer. This "mill-out joint" should be twice the length of the packer to fully accommodate the stinger when the shoe or mill has cut through the entire packer. If the mill-out joint is too short, the packer will probably be lost on the trip out of the hole, as the reciprocation of the work string in removing the slips will cause the release device to seat in the packer bore and the catcher to become released.

If no provision has been made for a mill-out joint, a J-type mandrel must be run in the washpipe above the shoe and a spear fitted to it of a size to engage in the packer bore. The grapple on the spear may be pinned in the catching position with a small brass shear pin. At no other time should spear grapples be pinned in the catching position.

Rotation of the carbide shoe or mill should follow the same general rules of cutting in other situations. Occasionally packers or parts of them will start to rotate and cutting will stop. When this occurs, it is necessary to shut the pump off and dry drill or spud the shoe or mill enough to foul the packer, so that it does not turn.

Chapter 14
Fishing Coiled Tubing

Coiled tubing presents unique problems in fishing somewhat similar to wireline. When coiled tubing parts, it is in tension and tends to corkscrew. It is also necked down for several inches at the point it parts. If the standard overshot is used, pressure or a downward force is necessary to push the pipe up through the grapple. This causes more corkscrewing of the fish and creates additional problems.

Continuous overshots have been designed especially for fishing coiled tubing (Figure 14-1). They are made with a long bowl and are run on tubing small enough to go inside the production tubing and with large enough I.D. to swallow the coiled tubing. The slips or grapples in the continuous overshots are split and offer little resistance to the fish as they go over it. Guide springs are provided to center the fish. When the work string is picked up, the guide springs tend to move the grapple down on the tapers and, in turn, engage the coiled tubing. This type of overshot can sometimes be worked over several hundred feet of coiled tubing. A pull on the work string tends to straighten out the fish so that even more can be swallowed. When sufficient pull is exerted, the coiled tubing can be pulled apart or all of it can be pulled from the well.

When coiled tubing is parted in casing, a mouse-trap-type tool (Figure 21-1) must be run and worked over a length of the tubing so that it can be pulled in two or removed from the well.

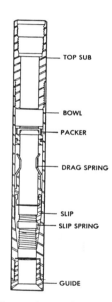

Figure 14-1. Continuous overshot for fishing coiled tubing. (Courtesy of Fishing Tools, Inc.)

Chapter 15
Fishing in Cavities

Frequently, when drilling, the pipe will part in a washed out section of the wellbore and the fish will not be centered in the hole. The straight overshot tool string may bypass the top of the fish and touch the pipe and take weight below the top. If this occurs, rotation slows and the cut-lip guide builds up slight torque and then jumps off. It may be impossible to engage the top of the fish with the tool string.

Bent Joints

A joint of pipe slightly bent just above the pin end and run just above the overshot will cause the tool to hang at an angle, and by rotating it near the top of the fish, it may be possible to engage the fish. This set-up is usually the first choice since it is simple and readily available on location. Some operators run a jet sub just above the overshot. This causes some of the pump pressure to be exerted against the wall of the hole which kicks the tools to the far side. This is advisable only on limited occasions as the jet washes the sidewall causing the filter cake to be washed off and eroding the hole.

Some subs have been cut so that the two ends are at a slight angle to each other. These are referred to as "bent subs," "crooked subs," "offset subs," and "angle subs." They are used instead of the drill pipe joint that is bent.

If the bent joint alone is not sufficient to catch the top of the fish, a wall-hook guide (Figure 15-1) can be substituted for the cut-lip guide on the bottom of the overshot. This guide is made so that it catches the pipe below the top and torque can be built up and held. By slowly picking up the work string, the fish is worked into the opening and fed into the overshot bowl (Figure 15-2).

Care should always be used in running a wall-hook guide, as excessive weight or torque can break the "hand" off. A considerable moment of force is built up when torque is applied.

Knuckle Joints

Figure 15-1. Wall-hook for overshot. (Courtesy of Bowen Tools.)

If the foregoing procedure will not catch the fish, then a knuckle joint can be added to the string just above the overshot fitted with the wall-hook guide. The knuckle joint can be compared to a hinge; it moves in one plane only. The entire string consisting of the wall-hook guide, overshot, and knuckle joint (Figure 15-3) is made up together on the catwalk and checked to be certain that the wall-hook moves out with the opening facing forward when the string is rotated to the right. Shims are provided so that the wall-hook opening can be adjusted until it is in the proper plane. The knuckle joint merely swings free as it is run in the hole. Pump pressure against the restriction plug (Figure 15-4) causes the overshot to be kicked out at a $7\frac{1}{2}°$ angle. With pump pressure holding the assembly out at an angle, the string is rotated to engage the fish. The restriction plug may be placed in the knuckle joint before running or it may be pumped down the pipe to its seat.

A knuckle joint, by design, is weak since it is a hinge. It will withstand very little jarring. If the fish is engaged with the foregoing set-up and it cannot be pulled, the restriction plug can be fished out with a small overshot run on the measuring line. This provides full opening of the tools, and a free point and string shot or cutter can be run and the pipe parted below the tools

94 Oilwell Fishing Operations

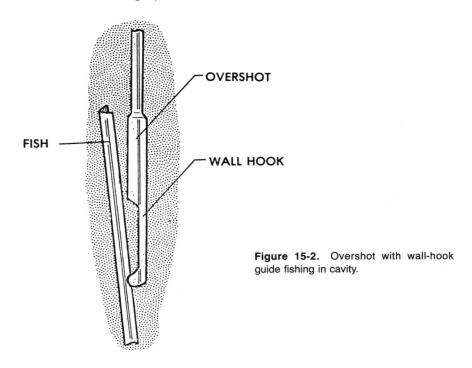

Figure 15-2. Overshot with wall-hook guide fishing in cavity.

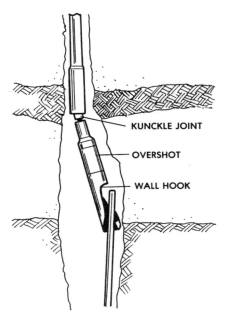

Figure 15-3. Fishing with wall hook and knuckle joint in cavity.

and in a section of hole that is more in gauge. This freed portion of the fish can be removed and the overshot run back in without the knuckle joint so that the fish can be jarred.

It is possible to get a very large sweep with the knuckle joint and overshot by adding extensions between the two. This set-up has been used to sweep large cavities that have been created in old wells by the use of nitroglycerin shots.

Induction Logs

Induction logging is a method in which the conductivity (the opposite of resistivity) of the formation is measured. Induced currents are used without the help of contact electrodes. A focused logging method requires no current flow from the tool into the formation, therefore, this method can be used in empty holes or in holes containing oil-base, fresh water, or other types of drilling fluid that are nonconductive.

Based on the preceding description of induction logging, one can easily see that this procedure can be used to locate the top of a fish in an open hole if it is impossible to determine its depth with the usual fishing tools. Frequently, when one is fishing in an open hole with washed out cavities in it, the top of the fish may be some distance from the center of the borehole, and because of corkscrewing of the pipe setting in compression, the top of the string may be difficult to otherwise find and identify.

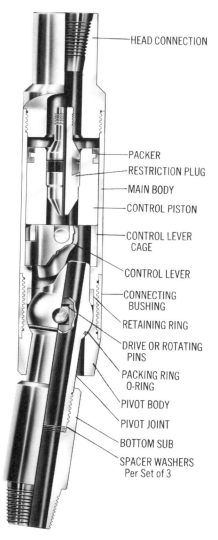

Figure 15-4. Knuckle joint angled at full 7½° by pump pressure on restriction plug. (Courtesy of Bowen Tools.)

Chapter 16
Sidetracking Junk

Under certain conditions, it is more practical and economical to sidetrack junk lodged in the wellbore than to remove it. In some areas, sidetracking in open-hole drilling operations is rather common. It has been found, from experience, that the average cost is less and the operation of the wells under producing conditions is not adversely affected by the deviation.

Although sidetracking through casing is not as prevalent, it is usually used when fishing the obstruction out is impractical or when an old well is reentered using directional drilling techniques to complete the new well in a different area. To sidetrack through casing, set a whipstock and cut the window at a depth where there is good cement bond. To determine where this is, run a cement bond log at the point of the proposed sidetrack. If there is no firm cement bond, the casing will move during the cutting and a poor job will result. If the bond is not sufficient, then the casing should be perforated and cement squeezed around it. An alternate depth might be selected where an adequate bond was found if this is practical and in keeping with the overall plan.

Casing whipstocks have been used for years with various methods of setting and holding in the casing. Various triggers have been used to trip the slips when pulled up into a coupling recess, as well as setting a cement plug to hold the whipstock in place.

The most modern method of setting the whipstock is to set it into a special keyway made into a permanent packer (Figure 16-1). By setting the packer on tubing or drill pipe, it may be oriented so that the window to be cut in the casing will be in the desired direction. This orienting should be checked after the packer has been set. When this is completed, the whipstock with a special keyway slot in a stinger on the bottom of the assembly is run on the drill pipe. Adjustments may be made in the location of the keyway slot to accommodate any variation that occurred in setting the packer. The whipstock is run pinned to the starting mill with a shear pin. The mill is made with a stinger which serves two purposes. The stinger holds the whipstock, and it also guides the starting mill by keeping it inside the casing, cutting a long window instead of merely cutting a hole, and going outside. Most shear pins are made to shear with approximately 10,000 lb–15,000 lb of weight after the whipstock has been guided into place above the packer. Once the pin has been sheared, rotation and circulation can begin, and the first phase of cutting the window is accomplished. Mills used for this purpose are made with both carbide and diamond cutting materials. Most operators do not run any drill collars with the starting mill, as it is desirable for it to follow the taper of the whipstock.

The window in the casing is completed with another mill. Cutting material, either diamonds or carbide, is dressed on the bottom as well as the sides. This mill is usually designed with a concave inset on the bottom so that it will ride down the casing. A single drill collar also helps to hold it into the casing so that a long window can be cut and the approximate taper of the whipstock followed. Some additional hole in the formation should be cut with this assembly so that the new hole is guided away from the old hole. In subsequent drilling of the new hole, a "watermelon" mill (Figure 11-6) is frequently run one or two joints above the bit to trim away any burrs and to help open up the window in the casing.

Figure 16-1. Casing whipstock set with key in slot of permanent packer.

Chapter 17
Section Mills

Section mills (Figure 17-1) are used to mill away complete sections of casing. Downhole section milling of casing is generally done for one of the following reasons:

- To mill away the perforated zone in an oil string, permitting under-reaming and gravel packing or completion in open hole.
- To mill away a section of casing to permit a sidetracking operation. With a downhole motor, the new hole may be started in any direction since a full 360° opening is provided.
- To mill away a loose joint of surface pipe.
- To cut pipe downhole for any purpose such as abandonment.
- To blank off a storage zone in a reservoir by removing a section above and below and then squeeze cementing the storage zone.

Ordinarily, in sidetracking operations, 25–30 feet of casing is milled up. This is sufficient for easy exit from the pipe, and it is also a convenient interval between couplings. Casing couplings or collars may be located by extending the blades with light pressure as the tool is lowered in the well. Less weight on the weight indicator identifies the blades in the recess of the couplings.

Section mills are made with cutting blades, dressed with tungsten carbide, and operated by pump pressure. A piston in the tool body is moved

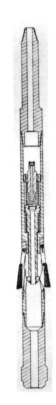

Figure 17-1. Section mill. (Courtesy of Petco Fishing & Rental Tools.)

in a cylinder by the pump pressure and in turn forces the blades out against the pipe. When the pipe is cut through, the blades are extended through the gap and then as weight is applied, milling the casing is accomplished. Drill collars are always run above the section mill to stabilize it and to afford the operator control of the weight. Sufficient circulation of a viscous fluid is necessary to remove all cuttings. A large amount of steel is removed; therefore, there is an abnormal amount of steel cuttings to be circulated out, screened, and removed from the mud. A ditch magnet in the return line will help to pick up all the steel particles.

Chapter 18
Repair of Casing Failures

Casing Leaks

Leaks in casing may be caused by many reasons such as burst with excessive internal pressure, improper make-up with subsequent leak at the threads, corrosion holes, eroded holes due to leaks in tubing, and perforations that may no longer be needed or desired.

The type of leak and its extent will probably dictate the method used to repair the casing. First, the exact location of the damage must be determined. This is usually done by pressuring between a bridge plug and a retrievable packer. The packer is moved until the hole or leak can be accurately determined, then a decision made as to the best method of repair.

Squeeze cementing is probably the most common method of sealing leaks in casing. Cement is pumped out through the leak and allowed to set up and the repair is then tested. Sometimes it is necessary to stage the cement job and to leave some cement in the casing under pressure until it sets. Then it is necessary to drill out the cement plug before the repair can be tested.

Liners may be set to blank off a section of casing which leaks. This can be a liner that is set all the way to the bottom of the hole and hung in the casing above the leak in the same manner as in an open hole. Liner hangers may be used that incorporate a packer to seal at the top of the liner between it and the casing, while other installations, such as the seal, depend entirely on cement. Liners restrict the size of the casing, which limits certain operations and equipment in the future. This may be a factor which will rule out the use of a liner. If the leak is high in the wellbore and it is not practical or economical to set a liner to the bottom of the well, a "scab" liner can be set across a short section of the casing including the leak. In this application, the liner is set on slips of a liner hanger with packer. The top is equipped with a setting sleeve and a liner packer with hold-down slips. This installation will pack off the area with the packers top and bottom, thereby isolating the leak. The disadvantage of this solution is that there is a restricted, smaller diameter section of casing in the well with the larger-diameter casing below.

If the economical considerations will warrant the cost of the job, faulty casing, including those joints with leaks or excessive corrosion, may be removed by cutting the casing below the damaged pipe, removing it, and running new casing with a casing patch or bowl to tie it back to the pipe left in the well. This procedure entails location of the lowest leak or the running of a casing inspection log to determine the lowest depth of pipe that has deteriorated and then cutting the casing with a mechanical internal cutter (Figure 6-7) run on a work string of tubing or drill pipe. The mechanical cutter is run to the depth desired and rotated to the right. which releases the slips. Then as slight weight is applied to the string, the knives are fed out on the tapered blocks and continued rotation cuts the casing. The new casing is made up with the casing bowl or patch on bottom and run in to the top of the cut-off portion of casing. The patch is engaged in much the same manner as an overshot by slow rotation to the right as weight is slacked off. The seal is effected, the grapple engages the pipe, and then the casing can be hung with the proper weight, as in a new string.

Casing patches or bowls (Figure 18-1) are made in several styles. There are two primary types of seals: neoprene and lead. The neoprene seal is rated at a higher pressure while the lead is believed to be more resistant to corrosion.

There are variations in the design of casing patches for different applications. One style permits the displacement of cement outside the pipe and through the patch prior to its sealing off (Figure 18-2). A casing patch is also manufactured with a long oversize extension on top for the salvage of pipe which has stuck before landing on a sub-sea wellhead. In this application, the casing is cut off with a mechanical internal cutter above the stuck point; the patch is run with a shorter string of pipe above it and landed over the casing that was cut off. After landing in the sub-sea wellhead, a

Figure 18-1. Casing patch bowl. (Courtesy of Bowen Tools.)

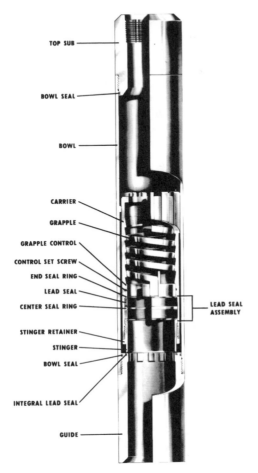

Figure 18-2. Cementing-type casing patch or bowl. (Courtesy of Bowen Tools.)

spear can be run inside the patch, engaged in the lower string of casing, and pulled up to the proper tension.

Casing patches have an excellent advantage in that they are full opening and full strength. They normally have a slightly larger O.D. than the standard coupling for the size of casing. Once they have been engaged and the casing landed, they should be pressure-tested to ensure that a seal has been accomplished. If an initial test is satisfactory, it is doubtful that there will ever be a problem of leakage, even after many years.

Casing Back-Off

If casing is badly pitted or has leaks at shallow depths, it is possible to back it off with special tools and screw back in with a die nipple (or collar) and new casing to the surface. This procedure was perfected in East Texas to economically repair the many casing leaks experienced above 600 ft. Depending on the depth of the surface pipe, it is possible to repair casing by this method down to 3,000 ft. The method is more practical, however, to approximately 1,000 ft.

Using left-hand tools (i.e., spear and drill pipe, and high torque power tongs), the casing is unscrewed approximately three rounds at the top joint or within a few joints of the surface. The spear is then moved down to a lower joint which is unscrewed the same amount. This procedure makes up the first thread that was broken. The same procedure is continued down the hole until the pipe is unscrewed below the lowest hole or defect. At this time the entire string can then be completely backed off and pulled from the well. The success of this procedure is due to the limited friction created by the rotation of a single joint at one time. Note that as the lower joints are broken and unscrewed, joints above that point are being made up, and therefore no pipe will ever completely unscrew and fall loose in the hole.

A line-up joint made to centralize the pipe is used above the spear to center the tool string and to prevent cross-threading the casing and die nipple (die collar) when it is screwed back in the pipe left in the well.

This method has been used quite successfully for thousands of wells. It is only limited by the depth of the casing leak and any pressure potential, since the wellbore is left open during a portion of the operation. There is considerable cost saving since there is no need of an expensive tie-back connection such as a casing patch bowl.

Stressed Steel Liner Casing Patch

This type of patch consists of a mild steel tube of approximately .125 in. in wall thickness formed inside the casing string to seal off perforations or other small leaks. The tube is first annealed, and then its diameter is decreased by pulling it through a series of rollers to corrugate it. The tube is again annealed to relieve the stresses set up when it is corrugated. The diameter of the tube selected for each size and weight of casing is most important as it must be stressed beyond yield as it is formed in the casing. Normally, the steel liner is about two percent larger in circumference than the inside circumference of the casing in which it is placed. The liner tube is covered with a layer of fiberglass, which acts as a carrying medium for an epoxy and as a gasket. Figure 18-3 shows a cross section of the tube before and after corrugating, with the glass fabric cemented on and the patch installed in the casing.

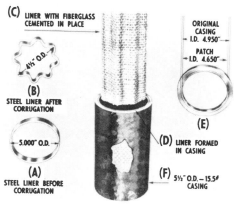

Figure 18-3. Steps in preparation and setting of stressed steel liner. (Courtesy of Petco Fishing and Rental Tools.)

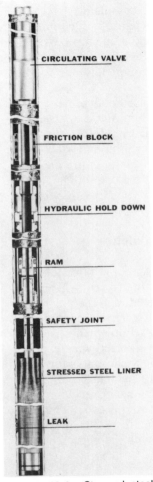

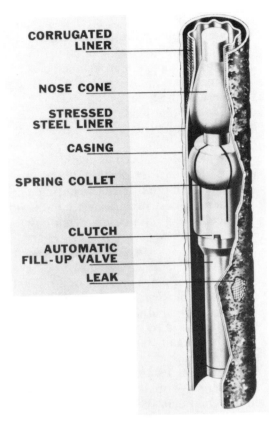

Figure 18-4. Stressed steel liner placement tool. (Courtesy of Petco Fishing & Rental Tools.)

Figure 18-5. Stressed steel liner setting tool. (Courtesy of Petco Fishing & Rental Tools.)

The placement tool (Figure 18-4) is made up of a hydraulic ram which pulls a two-stage setting tool through the corrugated liner. Just prior to running, the fiberglass is coated with an epoxy formulated for the running time and well temperature. The setting tool (Figure 18-5) is made up with a solid cone as the first stage and a spring collet as the second stage. The mild steel tube yields and forms a round, tightly fitted sleeve inside the casing. Since the original size of the tube was greater than the inside diameter of the casing, the liner is left in a hoop compressive stress.

Chapter 19
Collapsed Casing

Collapsed casing can be a very provoking problem because it is not always easy to determine the severity and the extent of the damage. Because of this, repair operations should be carefully planned and executed. First, determine insofar as possible, the length of the collapse by means of the allowable movement of the tubing and measurements correlated by free-point instruments. This determination is important because it is imperative that the tubing be cut above and below the collapsed section. Electric wireline cuts should be made leaving enough free pipe above the collapse for an easy overshot catch and enough below the collapse so that in swaging or rolling the casing, the tubing left in the well will not become an obstruction. The section of tubing that is cut top and bottom should be caught with an overshot and jarred out with the usual string of bumper jar, oil jar, drill collars, and intensifier.

After the section of tubing is removed from the wellbore, an impression block should be very carefully run in to take an impression of the collapsed pipe. The impression block should be measured out of the hole as the depth may be quite critical. If the collapse runs up the hole from the most severely damaged point, then the impression block will be depressed on the side and indicate that it has been wedged into a taper (Figure 19-1). This type of collapse does not present any special problem other than its severity. If the collapse occurs at a coupling and extends down the wellbore, the impression block will be marked only on the bottom where it has been pushed down on the end of the collapsed joint (Figure 19-2). This is an indication that the configuration downhole is similar to a whipstock, and any tools run in the hole may have a tendency to go outside the pipe. Carbide mills should be avoided in repairing casing unless it is impossible to reform the pipe. Tapered mills tend to "walk" and will invariably follow the path of least resistance and go outside the pipe in the situation just described.

106 Oilwell Fishing Operations

Casing swages, or swage mandrels (Figure 19-3) are heavy tapered cones which can be driven down through the collapse and jarred back out. It is usually necessary to run several sizes in sequence as the pipe must be swaged out in small increments, sometimes as little as 1/4 in. Most collapsed casing can be swaged out to approximately 1/8 in. below the drift diameter.

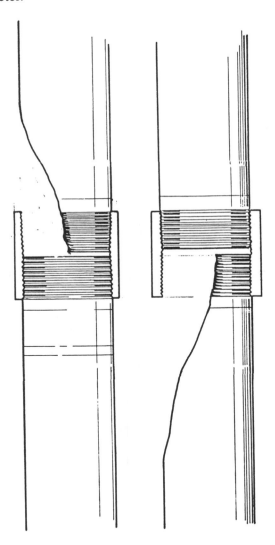

Figure 19-1. Pipe collapsed upward from coupling.

Figure 19-2. Pipe collapsed downward from coupling.

Collapsed Casing   107

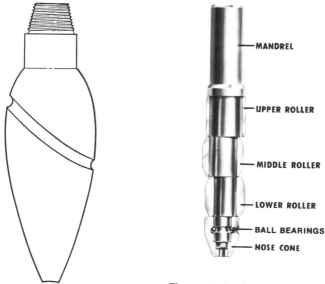

Figure 19-3. Casing swage mandrel.

Figure 19-4. Casing roller. (Courtesy of Bowen Tools.)

Casing rollers were first made as adaptations of the swage mandrel by merely adding a series of rollers, but they have been improved and made more sturdy (Figure 19-4). The operation of swaging or rolling is rather sever, regardless of which method is chosen. One should always run the jars and drill collars in either procedure, as they become hung and wedged in the collapsed section and must be jarred loose.

Once the casing is opened up sufficiently for normal operations, it must be reinforced in some manner prior to pumping it down or exposing it to the external pressure which caused the collapse initially. This can be done by squeeze cementing or setting a liner through the section.

Chapter 20
Fishing in High Angle Deviated and Horizontal Wells

Obviously, successful fishing in highly deviated wellbores is limited compared with the results possible in relatively straight wells. There is now sufficient experience in drilling highly deviated wells that an unusually high ratio of fishing is not required.

Most tools that are used in straight holes can be successfully run in highly deviated wells, with the exception of washover pipe. Because this pipe is large and not very flexible, any section must be extremely short in order to pass the deviated portion of the hole. Jars, overshots, magnets, and junk baskets may be successfully used.

When high angle holes have been drilled by rotating the drill pipe, a trough is usually formed which is smaller in diameter than the drilled portion of the hole. This should be considered when fishing with an overshot or similar tool, as the fish will lie in the trough or smaller section,

Chapter 21
Miscellaneous Tools

Mouse Traps

Mouse trap is the term commonly used to describe a catching tool that has a movable slip so that a variable catch can be made. Ordinarily it does not release and is therefore limited in use. The advantage is, of course, that it will catch fish that vary in size or that are of an unknown size. Mouse traps are most commonly used to catch sucker rods.

Larger versions of tools made with the mouse-trap principle are used to catch such fish as corkscrewed rods in casing, tubular fish, such as mud anchors, and corroded pipe or mashed pipe where the diameter is not consistent or standard. One such tool is the Clulow socket (Figure 21-1) which was originally manufactured for cable tool use. It consists of a bowl of appropriate size for the casing and two tracks running from top to bottom of the bowl and set at an angle. A suitable slip is fitted in the track. The slip is not anchored but free to slide up and down the track. A fish pushes the slip up the slanted track until sufficient clearance is available for the fish to pass the slip. When this is done, the slip falls down behind the fish and wedges it in the bowl. The fish can then be retrieved or pulled in two.

Reversing Tools

Reversing tools are used to unscrew and recover sections of right-hand pipe or tools that are stuck or lodged in the casing. The reversing tool (Figure 21-2) is used with a right-hand work string and by means of planetary gears and an anchoring system the right-hand rotation is converted to left-hand rotation below the tool. Left-hand threads on tools and pipe must be used below the reversing tool. The gear ratio of the planetary gears is two to one, so it is possible to get twice the torque below the tool as compared with the right-hand torque at the top of the tool. Reversing

110 Oilwell Fishing Operations

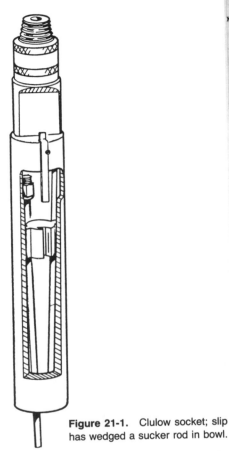

Figure 21-1. Clulow socket; slip has wedged a sucker rod in bowl.

Figure 21-2. Reversing tool. (Courtesy of Houston Engineers, Inc.)

tools have restricted internal diameters but the opening is usually large enough to accommodate a string shot. Since the reversing tool must be anchored in the casing, it is never run in open hole.

Ditch Magnets

The ditch magnet is a permanent magnet, usually two or three feet long, installed in the drilling mud return line between the shale shaker and the mud pits. Practically all ferrous metal particles are caught in this manner and kept out of the system where they cause undue wear on the circulating equipment and create further problems downhole.

Ditch magnets are used almost always on milling jobs, as it is impossible for the shaker screens to pick up all of the cuttings. Besides the advantage of keeping the abrasive particles out of the system, it is sometimes helpful to accumulate the cuttings and by weight determine how much cutting has been accomplished downhole.

Mud Motors

The positive displacement mud motor (Figure 21-3) has been used very effectively to find the top of a fish, particularly casing that has been cut or shot off and may have leaned over in the hole so that it cannot be caught or entered with conventional tools.

The motor, equipped with proper bit or other tools, is run just below a bent sub or crooked joint of pipe so that it is thrust out from center. By rotating the pipe very little, a large cavity can be covered and the fish entered. Because the pipe does not have to be rotated to turn the bit or tool, once the fish is found, it can usually be entered, cleaned out, and brought back to a more centered position. This set-up can also be used effectively for locating and retrieving fish that are lost in deviated holes.

Impression Blocks

Impression blocks (Figure 21-4) are used to secure an imprint of a fish to determine its size and configuration so that a suitable tool or procedure can be selected. They are usually made of lead poured on a mandrel with sufficient wickers or barbs so that the lead will not be pushed off when weight is applied. Impression blocks have been made of many soft materials such as coal tar, soap, and wood. Lead is ordinarily used since it has high heat resistance and is not resilient. The impression block is run on pipe or it may be run on a sand line with a stand of pipe to stabilize it and afford sufficient weight to form an identifiable impression. The interpretation of the marks on an impression block may be made easier if the "negative" imprint is changed to "positive" by making a mold of modeling clay or other soft material.

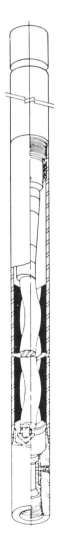

Figure 21-3. Positive displacement mud motor. (Courtesy of Teleco Oilfield Services, Inc.)

Figure 21-4. Lead impression blocks. (Courtesy of Petco Fishing & Rental Tools.)

Hydraulic Pull Tools

The hydraulic pull tool is a downhole hydraulic jack used for pulling liners, packers, and other equipment from a well without strain on the workstring or the derrick. It consists basically of three parts:

- A relief valve on top to open and close the tool to the annulus or casing (Figure 21-5).
- An anchor section consisting of hydraulic hold-down buttons which hold the tool firmly in the casing when pressure is applied (Figure 21-5).
- A five-cylinder hydraulic jack in which all of the cylinders are manifolded together and all pistons act on a common pull mandrel (Figure 21-6).

The tool is run with suitable catching tools on the bottom to engage the fish. The valve is then closed and hydraulic pressure applied to the work string. The hydraulic hold-down anchors the tool in the casing and the combined force of all five pistons acts to pull the mandrel up through its stroke. Since five cylinders are acting together, the pull ratio is quite

Miscellaneous Tools 113

Figure 21-5. Upper section of hydraulic pull tool (relief valve and anchor section). (Courtesy of Houston Engineers, Inc.)

Figure 21-6. Lower section of hydraulic pull tool (cylinder section). (Courtesy of Houston, Engineers, Inc.)

high. For the tool operating in 5½ in. casing, the ratio is 45 to 1, and since the tool is rated at 5,000 psi a pull of 225,000 lb is exerted on the fish. In the tool for 7-in. casing the ratio is 60 to 1, so a pull of 300,000 lb may be exerted. This force is exerted without any strain on the tubing or work string or the derrick, since the tool is anchored in the casing. It is never run in an open hole. The tool should be anchored several joints above the casing shoe, as there is an equal force pulling down on the casing. Catching tools run with the hydraulic pull tool are spaced out with suitable drill collars or other heavy pipe. A bumper sub is run above the catching tool so that the grapple may be bumped off the tapers if it is desired to release the tool. A bumper sub is also helpful to the operator because of its free travel. An internal cutter may also be run below a spear, and the fish (such as a liner) can be cut into shorter pieces for easier pulling. Safety joints are run below the pull tool and above the catching tool in case the grapple has become imbedded in the fish and it cannot be pulled.

The hydraulic pull tool is an excellent device for exerting unusual forces when light rigs and tubing are used in workover operations.

Tapered and Box Taps

Taps have only one advantage. They catch small or large holes or objects. Their catching size is variable. The disadvantage of taps is that they cannot usually be released. The principle of the tapered male (Figure 21-7) and female tap (Figure 21-8) is that they are self-threading with hardened threads and usually vertical grooves for the removal of cuttings.

Taps should not be run to catch pipe, drill collars, etc. that may be stuck, but they are practical for such small items as lift nubbins, bits,

Figure 21-7. Taper tap. (Courtesy of Gotco International, Inc.)

Figure 21-8. Box tap. (Courtesy of Gotco International, Inc.)

Miscellaneous Tools 115

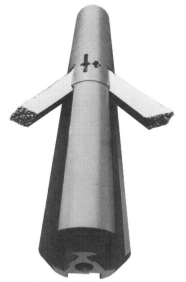

Figure 21-9. Marine pipe cutter. (Courtesy of Petco Fishing & Rental Tools.)

Figure 21-10. Marine swivel. (Courtesy of Petco Fishing & Rental Tools.)

balls of wireline, or any object that can be easily pulled and retrieved when caught.

Taps are usually made in a long configuration and taper down to almost a point. This is intended for them to be versatile according to size. However, they may "bottom up" through the hole in the fish before the threads engage. This should be determined where possible and the tap cut off before running. It may be cut at an angle or "mule shoed" with a cutting torch.

Marine Cutting Tools

When an offshore well is abandoned, regulations require that all strings of pipe be cut off below the mud line. Multistring cutters operated hydraulically are used for this purpose. The cutter consists of a cylindrical body with knives that are actuated by pump pressure on a piston in the body of the tool (Figure 21-9).

In order to maintain a constant depth from a floating drilling structure, the marine swivel (Figure 21-10) is landed in the wellhead. Bumper subs

Figure 21-11. Marine spear. (Courtesy of Petco Fishing & Rental Tools.)

Figure 21-12. Nonrotating stabilizer. (Courtesy of Petco Fishing & Rental Tools.)

or motion compensators above the tool absorb the wave action so that the cutter is at a fixed depth below the wellhead.

If all strings cannot be cut with one set-up, the smaller diameter strings are cut, the cutter retrieved, longer blades installed, and a compensating sub added to the string so that the cutter blades will be spaced at the previous cut. Cuts may be made in this manner up through 65-in. diameter pipe.

To retrieve the pipe that is cut, marine spears (Figure 21-11) are used. Different slips are installed on the body for the retrieval of the different sizes of casing.

To center and to stabilize the cutter, nonrotating swivels are used (Figure 21-12). Blades of the appropriate size are installed on the body to center the tool. The sleeve and blades do not rotate, but the mandrel rotates in the sleeve.

Glossary

Back off—To unscrew one threaded piece (as a section of pipe) from another.
Bailer—A long cylindrical container, fitted with a valve at its lower end, used to remove water, sand, mud, oil or junk and debris from a well.
Bent sub—A short cylindrical device installed in a drill stem between the bottom-most drill collar and a downhole mud motor. The purpose of the bent sub is to deflect the mud motor off vertical to drill a directional hole.
Boot basket—A tool run just above the bit or mill in the drill stem to catch small, nondrillable objects circulating in the annulus.
Box tap—A female tapered self-threading tool used to screw onto a fish externally for retrieval.
Bridge—An obstruction in the borehole, usually caused by the caving in of the wall of the borehole or by the intrusion of a large boulder.
Bumper jar (or *bumper sub*)—A percussion tool operated mechanically to deliver a heavy downward hammer blow to objects in the borehole.
Buoyancy—The apparent loss of weight of an object immersed in a fluid. If the object is floating, the immersed portion displaces a volume of fluid the weight of which is equal to the weight of the object.
Burning shoe—A type of rotary shoe designed to mill away metal; used in fishing operations.
Bushing—A pipe fitting which allows two pieces of pipe of different sizes to be connected together.
Cased hole—A wellbore in which casing has been run.
Catcher—A device fitted into a junk basket and acting as a trap door to retain the junk.
Collar—A coupling device used to join two lengths of pipe. A combination collar has different threads in each end.
Collar locator—A logging device for depth-correlation purposes, operated mechanically or magnetically to produce a log showing the location of each casing or tubing collar or coupling in a well. It provides an accurate way to measure depth in a well.
Completion fluid—A special drilling mud used when a well is being completed. It is selected not only for its ability to control formation pressure, but also for its properties that minimize formation damage.

Crooked hole—A wellbore that has deviated from the vertical. It usually occurs where there is a section of alternating hard and soft strata steeply inclined from the horizontal.

Die collar—A collar or coupling of tool steel, threaded internally, that is used to retrieve pipe from the well on fishing jobs; the female counterpart of a taper tap.

Dressing—A term used to describe the fitting together of all parts of a tool or the surfacing of a tool with particular materials such as "dressing" a mill with carbide.

Dutchman—A piece of tubular pipe broken or twisted off in a female connection. It may also continue on past the connection.

External cutter—A fishing tool containing metal-cutting knives that is lowered into the hole and over the outside of the length of pipe to cut it. The severed portion of the pipe can then be brought to the surface.

Fish—Any object in a well which obstructs drilling or operation; usually pipe or junk.

Flush-joint pipe—Pipe in which the outside diameter of the joint is the same as the outside diameter of the tube. Pipe may also be internally flush-joint.

Free point—The depth at which pipe is stuck, or more specifically the depth immediately above the point at which pipe is stuck.

Go devil—A device which is dropped or pumped down a borehole, usually through the drill pipe or tubing.

Grapple—The part of a catching tool (such as overshot or spear) that engages the fish.

Gyp—Gypsum.

Gypsum—A naturally occurring crystalline form of hydrous calcium sulfate.

Hydrostatic head—The pressure exerted by a body of liquid at rest. The hydrostatic head of fresh water is 0.433 per foot of height. Those of other liquids may be determined by comparing their specific gravities with the gravity of water.

Impression block—Tool made of a soft material such as lead or coal tar and used to secure an imprint of a fish.

Jar—A tool run in the string which imparts an impact either up or down.

Jar accelerator—A hydraulic tool used in conjunction with a jar and made up on the fishing string above the jar and drill collars to increase the impact.

Junk—Metal debris lost or left in a wellbore. It may be a bit, cones from a bit, hand tools, or any small object which is obstructing progress.

Junk basket—A cylindrical tool designed to retrieve junk or foreign objects loose in a wellbore.

Junk sub (also called *boot basket*)—A tool run just above the bit or mill in the drill stem to catch small, nondrillable objects circulating in the annulus.

Key seat—A channel or groove cut in the side of the hole parallel to the axis of the hole. Key seating results from the dragging of pipe on a sharp bend in the hole.

Kick—An entry of water, gas, oil, or other formation fluid into the wellbore. It occurs because the pressure exerted by the column of drilling fluid is not great enough to overcome the pressure exerted by the fluids in the formation drilled. If prompt action is not taken to control the kick or kill the well, a blowout will occur.

Knuckle joint—A hinged joint made up in the string above a fishing tool to allow it to be thrust out at an angle.

Liner—Any string of casing whose top is located below the surface. A liner may serve as the oil string, extending from the producing interval up to the next string of casing.

Long string—(1) The last string of casing set in a well. (2) The string of casing that is set through the producing zone, often called the oil string or production string.

Macaroni string—A string of tubing of very small diameter.

Magnet—A permanent magnet or electromagnet fitted into a tool body so that it may be run to retrieve relatively small ferrous metal junk.

Mandrel—A cylindrical bar, spindle, or shaft around which other parts are arranged or attached or that fits inside a cylinder or tube.

Measure in—To obtain an accurate measurement of the depth reached in a well by measuring the drill pipe or tubing as it is run into the well.

Measure out—To measure drill pipe or tubing as it is pulled from the hole, usually to determine the depth of the well or the depth to which the pipe or tubing was run.

Mill—A downhole tool with rough, sharp, extremely hard cutting surfaces for removing metal by cutting. Mills are run on drill pipe or tubing to cut up debris in the hole and to remove stuck portions of the drill stem or sections of casing for sidetracking. Also used as a verb to mean to use a mill to cut metal objects that must be removed from a well.

Milling shoe—See *Rotary Shoe* and *Burning Shoe*.

Mousetrap—A fishing tool used to recover a parted string of sucker rods or other tubular-type fish from a well.

Multiple completion—An arrangement for producing a well in which one wellbore penetrates two or more petroleum-bearing formations that lie one over the other. The tubing strings are suspended side by side in the production casing string, each a different length and each packed off to prevent the commingling of different reservoir fluids. Each reservoir is then produced through its own tubing string.

Necking—The tendency of a metal bar or pipe to taper to a reduced diameter at some point when subjected to excessive longitudinal stress.

Overpull—Pull on pipe over and beyond its weight in either air or fluid.

Overshot—An outside catch tool which goes over a tubular fish and catches it on the outside surface with a slip.

Pilot mill—A special mill that has a heavy, tubular extension below it called a pilot or stinger. The pilot, smaller in diameter than the mill, is designed to go inside drill pipe or tubing that is lost in the hole. It guides the mill to the top of the pipe and centers it over the pipe, thus preventing the mill from bypassing the pipe.

Pulling tool—A hydraulically operated tool that is run in above the fishing tool and anchored to the casing by slips. It exerts a strong upward pull on the fish by hydraulic power derived from fluid that is pumped down the fishing string.

Reverse circulate—To pump down the annulus and back up the work string (drill pipe or tubing). This is frequently used in workover in cased holes.

Rotary shoe—The cutting shoe fitted to the lower end of washover pipe and "dressed" with hard-surfaced teeth or tungsten carbide.

Safety joint—A threaded connection which has coarse threads or other special features which will cause it to unscrew before other connections in the string.

Sand line—A wire rope used on well-servicing rigs to operate a swab or bailer. It is usually 9/16-in. in diameter and several thousand feet long.

Sinker bar—A heavy weight or bar placed on or near a lightweight wireline tool. It provides weight so that the tool can be lowered into the well properly.

Spear—An inside catch tool which goes inside a tubular fish and catches it with a slip.

Squeeze cementing—The forcing of cement slurry by pressure to specified points in a well to cause seals at the points of squeeze. It is a secondary-cementing method, used to isolate a producing formation, seal off water, repair casing leaks, and so forth.

Stinger—Any cylindrical or tubular projection, relatively small in diameter, that extends below a downhole tool and helps to guide the tool to a designated spot (as in the center of a portion of stuck pipe).

String—The entire length of casing, tubing, or drill pipe run into a hole.

String shot (also called *Prima-Cord*)—An explosive line which when detonated imparts concussion to pipe causing it to unscrew or "back-off."

Sub (or *Substitute*)—A short section of pipe, tube, or drill collar with threads on both ends and used to connect two items having different threads; an adapter.

Surface pipe—The first string of casing set in a well after the conductor pipe, varying in length from a few hundred feet to several thousand. Some states require a minimum length to protect fresh-water sands.

Surfactant—A substance that affects the properties of the surface of a liquid or solid by concentrating on the surface layer. Reduces surface tension thereby causing fluid to penetrate and increase "wettability."

Swage (or *Swage mandrel*)—A tool used to straighten damaged or collapsed pipe in a well.

Taper tap—A male, tapered, self-threading tool to screw into a fish internally for retrieval.

Twist off—Of drill pipe or drill collars, to part or split primarily because of metal fatigue.

Underream—To enlarge the wellbore below the casing.

Wall hook—A device used in fishing for drill pipe. If the upper end of the lost pipe is leaning against the side of the wellbore, the wall hook centers it in the hole so that it may be recovered with an overshot, which is run on the fishing string and attached to the wall hook.

Washover pipe (or *Washpipe*)—Pipe of an appropriate size to go over a "fish" in an open hole or casing and wash out or drill out the obstruction so that the fish may be freed.

Bibliography

Adams, Neal, "How to Control Differential Pipe Sticking," *Petroleum Engineer*, Sept. 1977.

Askew, W. E., "Computerized Drilling Jar Placement," IADC/SPE 14746, Feb. 1986.

Brouse, Mike, "How to Handle Stuck Pipe and Fishing Problems," *World Oil*, Nov. 1982.

Brown, Michael C., "Fishing: What, Why and How Long," *Drilling Contractor*, Jan. 1985.

Fox, Fred K., "New Pipe Configuration Reduces Wall Sticking," *World Oil*, Dec. 1960.

Goins, W. C., "Better Understanding Prevents Tubular Buckling Problems," *World Oil*, Jan. 1980.

Grogan, Gene E., "How to Free Stuck Drill Pipe," *Oil and Gas Journal*, April 4, 1966.

Harrison, C. Glenn, "Fishing Decisions Under Uncertainty, *Journal Petroleum Technology*, Feb. 1982.

Huffstetler, J. T., "Decide—A Project Planning Tool," Nov. 12, 1970.

Kemp, Gore, "Field Results of the Stressed Steel Liner Casing Patch," *Journal of Petroleum Technology*, Feb. 1964.

Kemp, Gore, "Tungsten Carbide—The Material That Made Today's Mills Possible," *Drilling*, June 30, 1975.

Krol, David A., "Additives Cut Differential Pressure Sticking in Drillpipe," *Oil and Gas Journal*, June 4, 1984.

Love, T. E., "Stickiness Factor—A New Way of Looking at Stuck Pipe," IADC/SPE 11383.

McGhee, Ed, "Gulf Coast Drillers Whip the Wall-Sticking Problem," *Oil and Gas Journal*, Feb. 27, 1961.

Mondshine, T. C., "Drilling-Mud Lubricity," *Oil and Gas Journal*, Dec. 7, 1970.

Outmans, H. D., "Spot Fluid Quickly to Free Differentially Stuck Pipe," *Oil and Gas Journal*, July 15, 1974.

Pfleger, Kenneth A., "Stuck Drill Pipe? Surfactant May Save a Washover Job," *Oil and Gas Journal*, March 16, 1964.

Porter, E. W., "Fishing Is More Art Than Science," *Oil and Gas Journal*, Sept. 21, 1970.

Sartain, B. J., "Drillstem Tester Frees Stuck Pipe," *The Petroleum Engineer*, Oct. 1960.

Shryock, S. H., and Slagle, K. A., "Problems Related to Squeeze Cementing," *Journal of Petroleum Technology*, Aug. 1968.

Skeem, Marcus R., Friedman, Morton B., and Walker, Bruce H., "Drillstring Dynamics During Jar Operation," *Journal of Petroleum Technology*, Nov. 1979.

Wood, Thomas R., "U-Tube Method Frees Stuck Pipe," *Oil and Gas Journal*, March 31, 1975.

Index

Accelerator, 38
Adaptable tool, 66
Application of carbide, 69

Back-off, 23–24
　casing, 102
　with clean-out tool, 55
Back-off connector, 55
Bailer, hydrostatic, 64
　rotating, 64
Baskets, junk, 59–65
　boot basket, 62
　clulow socket, 63
　core type, 59
　friction sockets, 60
　poor boy, 62
　reverse circulation, 61
Bent sub, 92
Block, impression, 111–112
　collapsed casing and, 105
Booster jar, 38
Boot basket, 62–63
　mills and, 73
Bowl, casing, 101
Box taps, 82, 114
Bumper jar, 35–36
　cutting electric conduit, 86–87
　marine cutters and, 115
　pull tools and, 112
Buoyancy, 19

Cable guide, wireline fishing, 77
Carbide, tungsten, 68–76
　application, 69
　design, 69
　dressing, 69
　material, 68
Casing back-off, 102
Casing
　collapsed, 105–107
　leaks, 100
Casing patch, 101–104
　stressed steel liner, 103
Casing roller, 107
Casing swage, 106–107
Catching tools, 29–34
　overshots, 29–33
　spears, 32–34
Cementing
　casing patch, 101
　squeeze, 100
Center spear, 81
Chemical cut, 24–26
　electric submergible pump conduit, 85
Clulow socket, 63, 110
Coiled tubing, 91
Continuous overshots, 91
Core-type junk basket, 59
Curves, stretch, 17
Cut and strip, wireline fishing, 77

Cutters
  casing, 101
  chemical, 24–26
  electric submergible pump cable, 85
  external, 51–54
  internal, 27, 101
  jet, 26
  Kinley, 83
  marine, 115
  wireline, 83

Differential sticking, 9
Dimensions, recording, 5
Ditch magnet, 110
Drill collars, 39
Drill collar spear, 52–54
  preventing stripping job, 53
Drill stem test tool, 13

Electric submergible pumps, 85
Electric wireline tool, 85–86
Electromagnet, 58
External cutters, 51–54

Factors, probability, 2, 51
Fishing tool operator
  and bumper jar/constant weight, 36
  free point observation, 19
Free point instruments, 19
  with electric submergible pump, 85
Friction sockets, 60

Grab, wireline, 81–82
Guide, wall hook, 93

Head, retrieving, 66
Horizontal well fishing, 108
Hydraulic clean-out tools, 55–56
Hydraulic pull tool, 112–114
Hydrostatic bailer, 64

Impression block, 111–112
  collapsed casing and, 105
Induction log, 95
Intensifier, 38
Internal cutter, 27

Jars, 35–44
  accelerator, 38
  bumper jars, 35
  drilling jars, 42
  fishing retrievable packers, 86
  hydraulic jars, 36
  intensifier, 38
  jarring strings, 36–39
  oil jars, 36
  run as insurance, 3
  run with swages and rollers, 106
  surface jars, 41
Jet cut, 26
Jet sub, 92
Joint, bent, 92
  safety, 48
  unlatching, 55
Junk baskets, 59–65
  boot basket, 62
  clulow socket, 63
  core type, 59
  friction sockets, 60
  poor boy, 63
  reverse circulation, 61
Junk shots, 67

Key seat, 8
Knuckle joint, 93–95

Liners, 100
  "scab," 100
  stressed steel, 103–104
Logs
  casing inspection, 101
  cement bond, 96
  induction, 95
  stuck pipe, 20

Magnets, 57
  ditch, 110
  electromagnet, 57
  permanent, 57
Mandrel, swage, 107
Marine cutter, 115
Marine spear, 116
Mechanical cut, 27
Mill-out joint, 90

Index

Mills, carbide, 68–76
  collapsed casing and, 105
  flat bottom, 72
  pilot, 74
  section, 98
  string, 75
  tapered, 73
  watermelon, 75
Mouse traps, 109
Mud motors, 111

Nomograph, 15–16

Operator, fishing tool
  free point observation, 19
  and bumper jar use for constant weight, 36
Overshots, 29–32
  with cable guide, 77
  continuous, 91
  with knuckle joint, 93
  releasing, 31
  short catch, 32
  side door, 80

Packer retrieving tool, 89–90
Packers, permanent
  fishing, 89–90
  whipstock set in, 97
Packers, retrievable, 88
Parting pipe, 21–28
  by back-off, 22
  by chemical cut, 24
  by jet cut, 26
  by mechanical cut, 27
  by outside back-off, 24
Patch, casing, 101–104
  stressed steel liner, 103
Permanent packers, 89
Pilot mills, 74
Pipe sticking, 7–13
Pipe stretch, 14–17
Pipe, stuck
  blowout stuck, 8
  cemented, 8
  differentially stuck, 9–13
  key seat, 8
  log, 20
  lost circulation, 9
  mechanically, 7
  mud stuck, 7
  sloughed hole, 9
  undergage hole, 9
Pipe, washover, 45–56
  length and, 49
Poor boy junk baskets, 62
Probability factors, 2, 51
Pull tool, hydraulic, 112
Pumps, electric submergible, 85

Radioactive sources, 80
Reamer, blade, 75
Retrievable packers, 88
Retrieving tool, packer, 89
Reverse circulation junk basket, 59
Reversing tool, 109
Roller, casing, 107
Rope knife, 84
Rope socket, 6
Rope spear, 82
Rotary shoes, 49
  tooth type, 76
  carbide, 68–76
Rotating bailer, 64

Safety joint, 48
  unlatching type, 55
  pull tools and, 114
Section mills, 98
Shoe, rotary, 49
Shots, junk, 67
Side door overshot, 80
Sidetrack junk, 96–97
Spears, 32–34
  center, 81
  marine, 116
  pack-off, 34
  rope, 81
  stop sub, 34
  releasing, 33
  washpipe, 49–52
Spotting fluid, 12
Stabilizer, nonrotating, 116

Stressed steel liner, 103–104
Stretch, 14–18
  curve, 17
  formula, 15
  nomograph, 16
String mills, 75
String shot, 22
  after washover, 51
  through reversing tool, 110
Stuck pipe log, 20
Subs, 92
Submergible pumps, electric, 85
Surface jar, 41
Surge method, 11
Swage mandrel, 106

Tapered mills, 73
Taps
  box, 82–114
  tapered, 114
Test tool, drill stem, 13
Tools
  catching, 29–34
  hydraulic clean-out, 55–56
  hydraulic pull, 112

marine cutting, 115
reversing, 109
Torque, 23
  working down, 88
Tungsten carbide, 68–76
  application, 69
  design, 69
  dressing, 69
  material, 68

Unlatching joint, 55

Wall hook guide, 93
Washover pipe, 45–56
  length is important, 49
  specifications, 46–47
  threads, 48–49
Washpipe spear, 52–54
Whipstock, 96–97
Wireline fishing, 77–87
  cable guide, 77
  cut and strip, 77
  cutting line, 83
  grabs, 81–82
  parted line, 81
  side door overshot, 80